ENGINEERING MECHANICS

ENGINEERING MECHANICS

Manish Saxena

ANMOL PUBLICATIONS PVT. LTD.
NEW DELHI-110 002 (INDIA)

ANMOL PUBLICATIONS PVT. LTD.
Regd. Office: 4360/4, Ansari Road, Daryaganj,
New Delhi-110002 (India)
Tel.: 23278000, 23261597, 23286875, 23255577
Fax: 91-11-23280289
Email: anmolpub@gmail.com
Visit us at: www.anmolpublications.com

Branch Office: No. 1015, Ist Main Road, BSK IIIrd Stage
IIIrd Phase, IIIrd Block, Bangalore-560 085 (India)
Tel.: 080-41723429 • Fax: 080-26723604
Email: anmolpublicationsbangalore@gmail.com

Engineering Mechanics

First Edition, 2010

ISBN 978-81-261-4592-8

PRINTED IN INDIA

Printed at Balaji Offset, Delhi.

Contents

Preface

Textbook of Engineering Mechanics is the first of kind introducing structural and continuum mechanics in a comprehensive and consistent way. The current book presents all theoretical developments both in text and by means of an extensive set of figures. This same approach is used in the many examples, drawings and problems. Both formal and intuitive (engineering) arguments are used in parallel to derive the principles used, for instance in bending moment diagrams and shear force diagrams. A very important aspect of this book is the straightforward and consistent sign convention, based on the stress definitions of continuum mechanics.

The book is suitable for Civil, Mechanical, Naval, Aerospace engineers and Architectural and Industrial designers; a reference work for advanced students, engineers and researchers alike

Author

Chapter 1

Concept of Mechanics

PRINCIPLES OF MECHANICS

Having decided on our topic to model, the first stage is to recap on the principles that surround the mechanics that will govern the behaviour of our experiments. The fundamental "quantity of motion" of an object is the product of its mass and velocity. This quantity of motion is called momentum.

momentum = mass × velocity

Conservation of momentum: If no external forces act during a collision, the total momentum of the system is conserved. The law applies equally to the motion of subatomic particles as it does to the motion of planets. For example, rocket propulsion is just one practical application of this law.

NEWTON'S LAWS OF MOTION

Newtons First Law: A body will continue in its state of rest, or of uniform motion in a straight line, unless an external force is applied to it.

Newtons Third Law: When two bodies, A and B, interact, the force that A exerts on B is equal in magnitude and opposite in direction to the force that B exerts on A.

LAW OF RESTITUTION

Newtons experimental evidence showed that for two particular colliding particles, the separation speed was always the same fraction of the approach speed.

This implies that separation speed = e x approach speed. The value of e affects the elasticity of the collision.

Vectors

A vector is a quantity that has both magnitude and direction. In our model we will use position vectors to keep track of the objects that we are using. A position vector, v, in our case will be thought of as a linear combination of the standard basis vectors (e.g. v = xi + yj).

Some Theory

Recall the equation F = ma, where F is the horizontal force of motion, m is the mass and a is acceleration.

From the standard equation of motion v = u + at, t represents time, a is acceleration, u is the initial velocit, and v is the final velocity, we can rearrange to obtain an equation for a.

Therefore,

$$a = (v-u)/t$$

And we can rewrite: F = ma = m(v – u)/t (equation 1)

Lets try to apply this initial principle to a basic collision problem. Imagine two billiard balls moving horizontally in space in such a manner that they collide. Assume the two balls move in the same straight line both before and after collision.

If, before the collision, both balls (A and B) are travelling at steady speeds UA and UB, what can we expect to happen to the overall system of momentum of the collision?

Well, the result of the collision will introduce a speed for both balls, VA and VB. There will be some force, F, acting for some time, t. We can approach a solution to this with the help of Newtons' Third Law, which gives rise to an equal and opposite reaction. Therefore, the force excerpted by ball A upon B must be exactly the same as, but in the opposite direction to, the force exerted by ball B upon A.

Therefore, we can rewrite equation 1 and substitute for masses of balls A and B: –f x t = mAVA – mAUA (equation 2 – negative as it is in the opposite direction)

$$f \times t = mBVB - mBUB$$

We can add equations 2 and 3 to get rid of the unknowns f and t:

$$mAVA - mAUA + mBVB - mBUB = 0.$$

Which we can further rearrange to give:

mAUA + mBUB = mAVA + mBVB

We can that the left-hand side of equation 4 is the total momentum of the system before the collision, and the right-hand side is the total momentum after the collision.

Therefore, we can say that the conservation of momentum applies to this problem, and we would expect the total momentum after the collision will be exactly equal to the momentum before the collision.

Collisions are classified as to whether or not kinetic energy is conserved. If it is, the collisions are called 'elastic' collisions, if not they are called 'inelastic' collisions. Now we have had a quick recap, we are ready to move on to the next stage, called elasticity

MOTION

In preparation for our discussion of Newton's laws, we will now look at how the motion of objects may be described. Three concepts - position, velocity, and acceleration - are introduced and defined.

POSITION

The first idea needed in describing the motion of an object is its position relative to some fixed reference point.

This entails two ideas: The distance the object is away from the reference point, and also the direction relative to that reference point. For example, one could say that Toronto is about 3000 km east of Vancouver. Position, as defined above, is said to be a vector quantity, because it has both a magnitude, or size, (the distance) and a direction. Sometimes we wish to talk about the position of an object relative to its starting point at some initial time. This is called the displacement of the object.

VELOCITY

For an object in motion the concept of velocity is important. There are two types of velocity that can be defined. Suppose an object moves from one point to another in a certain time interval. The average velocity during that time interval is defined as

Average velocity=change in position divided by the time

Note that, as with the position, the velocity is a vector: it has a magnitude and a direction associated with it. For example, if someone travels from Toronto to Vancouver by airplane in three hours, their average velocity for that trip is 3000 km divided by 3 hours, or 1000 km/hr, due east. Sometimes only the magnitude of the velocity is of interest. It is called the speed.

The average speed for the above trip from Toronto to Vancouver is 1000 km/hr. No direction need be specified. Note that objects moving with a large speed change their position quickly over time, whereas those moving with a small speed take longer to cover the same distance.

The average velocity only depends on the initial and final points of the travel - it knows nothing of the details of the motion between these points. For this reason one introduces the concept of instantaneous velocity, which applies to a single point, or time of the motion. The instantaneous velocity is defined as follows. Suppose one wants to define the instantaneous velocity at a certain point, say A along the journey.

One first calculates the average velocity between this point and an arbitrary nearby point, say B_1. The arrow in the Figure below pointing from A to B_1 indicates the direction of the average velocity in going between those two points. One then does the same, but for a point B_2 which is nearer to the initial point A of interest.

Then again for B_3 which is closer still, and so on. The instantaneous velocity is obtained by repeating this process until you get to a point B that is arbitrarily close to A.

As indicates, as the B gets closer and closer to A the direction of the velocity (indicated by the arrows) gets closer and closer to pointing along a line that is tangent to the trajectory at the point A (i.e. points along the trajectory). The instantaneous speed is correspondingly defined as the magnitude of the instantaneous velocity.

An example with which we are all familiar involves travel by car. Suppose we are driving between Winnipeg and Kenora. This road is more or less completely straight, so we don't have to worry about directions. After two hours we stop for a one hour lunch break, and then continue to drive for another hour.

At this stage, the odometre shows that we have driven 300 km, so that our average speed for that portion of the trip would be 300/4 = 75 km/hr.

The magnitude of our instantaneous speed, on the other hand is given by the speedometre reading, and can be very different from the average speed. For example we might be going at 100 km/hr or if we happen to be stopped at a traffic light, our instantaneous speed might be zero.

In effect, the odometre counts the total number of revolutions made by the tires (and multiplies by the circumference of the tire) in order to get the total distance traveled.

The speedometre instead counts the number of revolutions in a very short time interval, say half a second, in order to tell us how fast the car is going at that instant.

ACCELERATION

As we saw, acceleration is a very important concept in Newtonian physics. Just as the velocity involves a rate of change of position in time, theacceleration of an object describes the rate of change of velocity. Also just as with velocity, there are two types of acceleration.

The first - the average acceleration - involves the change in instantaneous velocity over a given time interval: Average acceleration=change in velocity over time.

Note that, from now on, when we say "velocity", we will mean the instantaneous velocity. As well as a size, acceleration also involves a direction: for example, a car traveling on a road that runs due north accelerates from rest at a stop sign to a speed of 30 m/s in 10 seconds. That car experienced an average acceleration during that time of 30/10 m/s^2 or 3 m/s^2 with direction due north.

In terms of km/hr, 30 m/s is:

$$\frac{30\text{m}}{1\text{s}} \times \frac{1\text{km}}{1000\text{m}} \times \frac{3600\text{s}}{1\text{hr}} = \frac{108\text{km}}{1\text{hr}}$$

If the driver of the same car notices a radar trap and slows down to the speed limit of 60 km/hr, or about 17 m/s in, say 5 seconds, then the car undergoes an average acceleration due south of 17/5 m/s^2 = 3.4 m/s^2. In this case, the car is slowing down, or decelerating. Objects with a large acceleration change their velocity quickly over time.

We can also define the instantaneous acceleration at a point A as the limit of the average acceleration between A and a nearby point B as the interval between Band A becomes zero:

The instantaneous acceleration at A is the average acceleration between A and B as B approaches A.

From now on, when we say "acceleration" we will mean the instantaneous acceleration.

GRAPHING MOTION

It is often useful to represent the motion of objects pictorially on a position vs time graph. Such a graph takes the form of a line. Each point on the line specifies the position (relative to the origin) of an object at a given instant in time. The simplest "motion" is that of a stationary object, whose velocity and acceleration are zero.

The next simplest motion, is that of an object going at constant velocity (hence zero acceleration). Its position vs time graph is again a straight line, but with non-zero slope. In fact the slope of the line (the rise over the run) equals the speed of the object.

The red line in Fig.2.2 shows the position time graph of an object moving at 2 m/s along the positive x-axis. Since it is a straight line, the slope can be calculated between any two points on the graph. If the object were moving with constant velocity along the negative x-axis, the line would be sloped in the opposite direction: the value of the x-coordinate would decrease as time increased. Free fall motion

One important elementary example of motion is that of free fall, which is the motion of an object only under the influence of gravity. It is a remarkable fact, first discovered over 300 years ago by Galileo and others, that objects in free fall motion descend at the same rate, *independent of their mass,* as long as the effects of air resistance can be neglected. This means that under free fall all objects have the same constant acceleration, which in the metric system is 9.8 m/s^2 at sea level, directed towards the centre of the earth.

One fairly dramatic illustration of this fact can be done by taking a fairly large text book (say the one for this course), and a piece of paper that is just smaller than the pages in the book.

If you hold the book in your left hand and the sheet of paper in your right, then drop them simultaneously, it is quite evident that the two do not fall towards the Earth at the same rate. This shows why for a long time people believed that objects fell at different rates.

However, if you now put the piece of paper on top of the text book, and drop them together, you will see quite dramatically that the paper and the book drop at precisely the same rate. In the second experiment, clearly the text book, which is less affected by air resistance, is shielding the piece of paper from the affects of air resistance so that both are essentially moving only under the influence of gravity. Consequently, as stated above, they accelerate at the same rate.

Graphs of how the position, velocity, and acceleration of objects under free fall motion change in time follow.

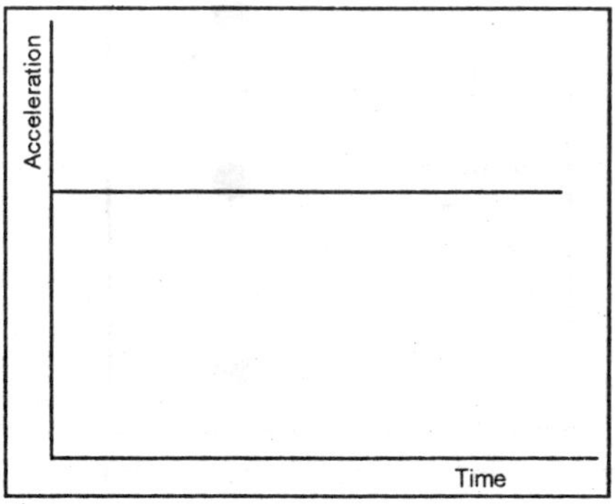

Fig. Free Fall Acceleration

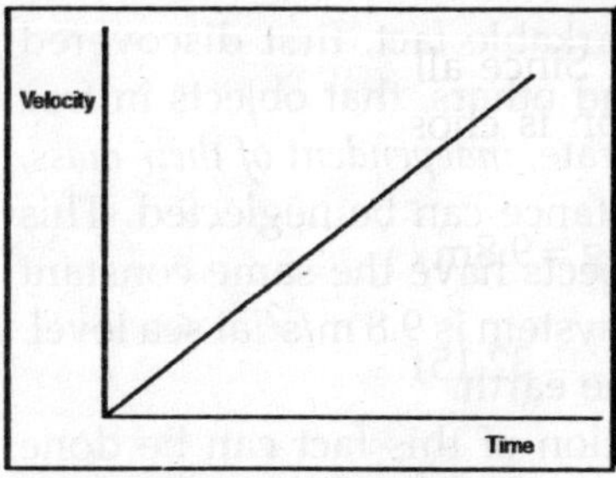

Fig. Free Fall Velocity

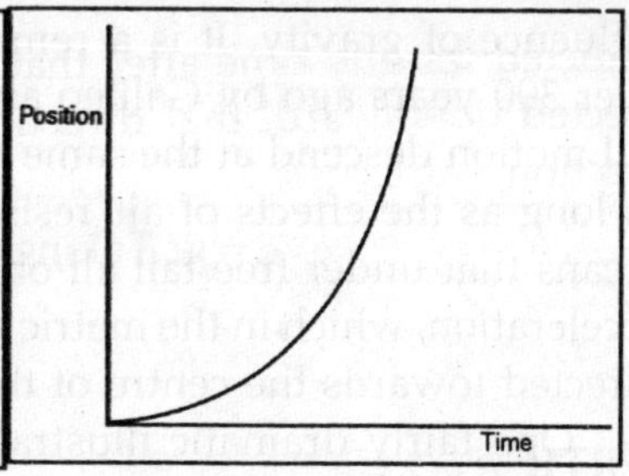

Fig. Free fall position.

TRAJECTORY

FREEFALL

In the absence of frictional drag, an object near the surface of the earth will fall with the constant acceleration of gravity g.

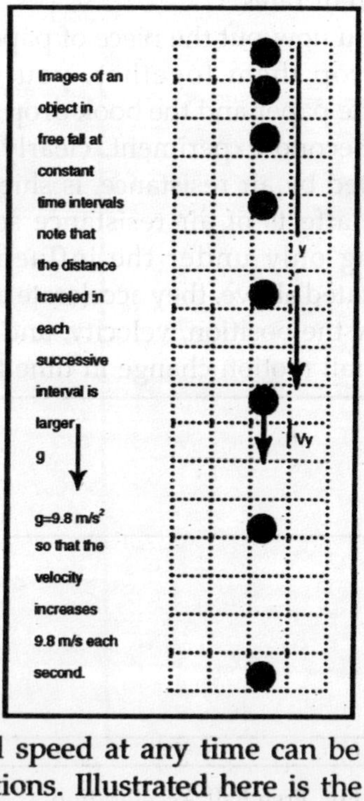

Position and speed at any time can be calculated from the motion equations. Illustrated here is the situation where an object is released from rest. It's position and speed can be

predicted for any time after that. Since all the quantities are directed downward, that direction is chosen as the positive direction in this case.

$$v_y = gt \quad \text{Taking } g = 9.8 m/s^2$$

$$y = \frac{1}{2}gt^2 \qquad = 32.15 ft/s^2$$

VERTICAL TRAJECTORY

Vertical motion under the influence of gravity can be described by the basic motion equations. Given the constant acceleration of gravity g, the position and speed at any time can be calculated from the motion equations:

$$v_y = v_{0y} - gt \quad \text{Taking } g = 9.8 m/s^2$$

$$y = v_{0y}t - \frac{1}{2}gt^2 \qquad = 32.15 ft/s^2$$

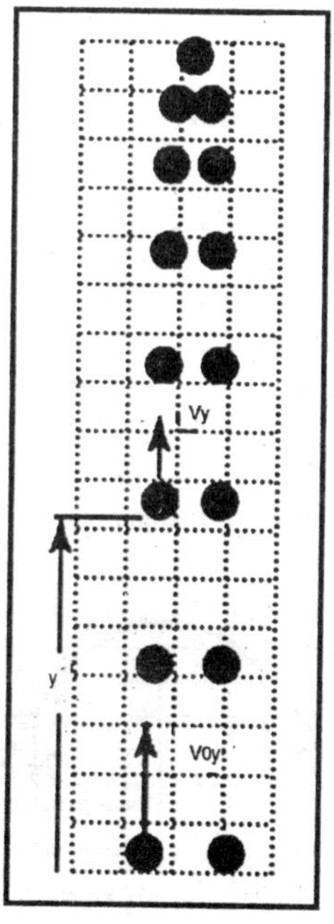

Horizontal Launch

All the parametres of a horizontal launch can be calculated

with the motion equations, assuming a downward acceleration of gravity of 9.8 m/s^2.

General Ballistic Trajectory

The motion of an object under the influence of gravity is determined completely by the acceleration of gravity, its launch speed, and launch angle provided air friction is negligible. The horizontal and vertical motions may be separated and described by the general motion equations for constant acceleration. The initial vector components of the velocity are used in the equations. The diagram shows trajectories with the same launch speed but different launch angles. Note that the 60 and 30 degree trajectories have the same range, as do any pair of launches at complementary angles. The launch at 45 degrees gives the maximum range.

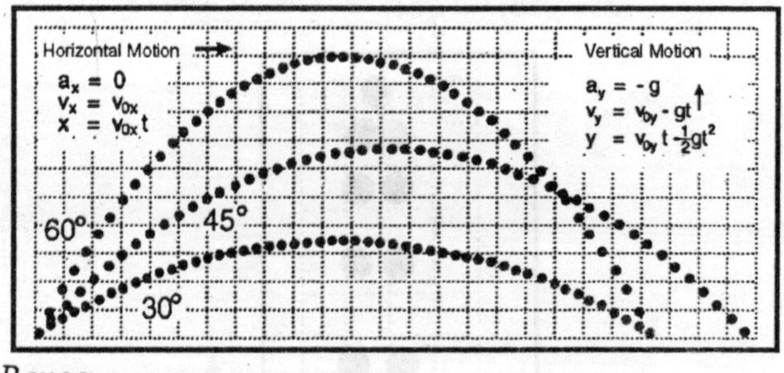

Range:

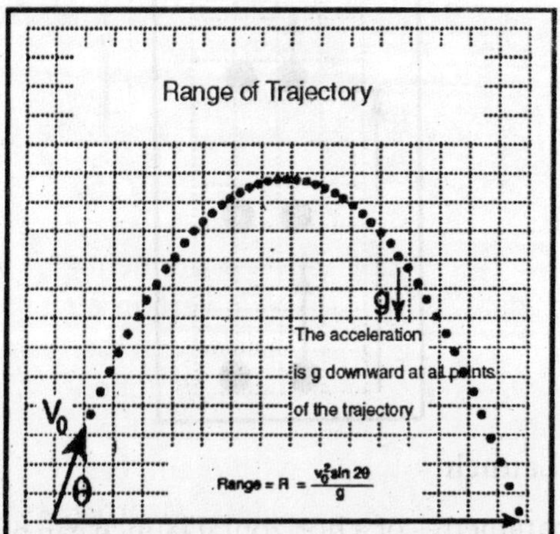

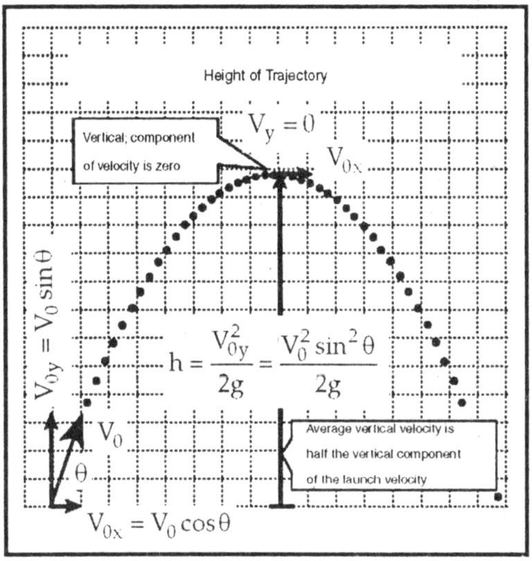

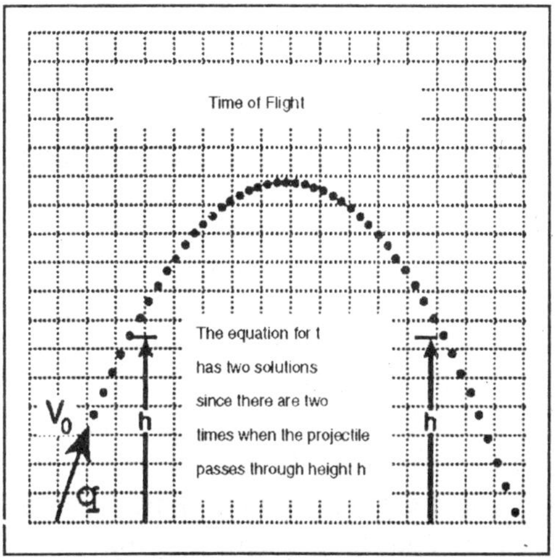

FORCE

So far you've studied the measurement of motion in some detail, but not the reasons why a certain object would move in a certain way. This chapter deals with the "why" questions. Aristotle's ideas about the causes of motion were completely wrong, just like all his other ideas about physical science, but it will be instructive to start with them, because they amount to a road map of modern students' incorrect preconceptions.

Aristotle thought he needed to explain both why motion occurs and why motion might change. Newton inherited from Galileo the important counter-Aristotelian idea that motion needs no explanation, that it is only *changes* in motion that require a physical cause. Aristotle's needlessly complex system gave three reasons for motion:

Natural motion, such as falling, came from the tendency of objects to go to their "natural" place, on the ground, and come to rest.

Voluntary motion was the type of motion exhibited by animals, which moved because they chose to. Forced motion occurred when an object was acted on by some other object that made it move.

CHANGES OF MOTION

In the Aristotelian theory, natural motion and voluntary motion are one-sided phenomena: the object causes its own motion. Forced motion is supposed to be a two-sided phenomenon, because one object imposes its "commands" on another. Where Aristotle conceived of some of the phenomena of motion as one-sided and others as two-sided, Newton realised that a change in motion was always a two-sided relationship of a force acting between two physical objects.

The one-sided "natural motion" description of falling makes a crucial omission. The acceleration of a falling object is not caused by its own "natural" tendencies but by an attractive force between it and the planet Earth. Moon rocks brought back to our planet do not "want" to fly back up to the moon because the moon is their "natural" place. They fall to the floor when you drop them, just like our homegrown rocks. As we'll discuss in more detail later in this course, gravitational forces are simply an attraction that occurs between any two physical objects. Minute gravitational forces can even be measured between human-scale objects in the laboratory.

The idea of natural motion also explains incorrectly why things come to rest. A basketball rolling across a beach slows to a stop because it is interacting with the sand via a frictional force, not because of its own desire to be at rest. If it was on a frictionless surface, it would never slow down. Many of Aristotle's mistakes stemmed from his failure to recognize friction as a force.

The concept of voluntary motion is equally flawed. You may have been a little uneasy about it from the start, because it assumes a clear distinction between living and nonliving things. Today, however, we are used to having the human body likened to a complex machine.

In the modern world-view, the border between the living and the inanimate is a fuzzy no-man's land inhabited by viruses, prions, and silicon chips. Furthermore, Aristotle's statement that you can take a step forward "because you choose to" inappropriately mixes two levels of explanation. At the physical level of explanation, the reason your body steps forward is because of a frictional force acting between your foot and the floor. If the floor was covered with a puddle of oil, no amount of "choosing to" would enable you to take a graceful stride forward.

MEASUREMENT OF FORCES

In the Aristotelian-scholastic tradition, the description of motion as natural, voluntary, or forced was only the broadest level of classification, like splitting animals into birds, reptiles, mammals, and amphibians.

There might be thousands of types of motion, each of which would follow its own rules. Newton's realization that all changes in motion were caused by two-sided interactions made it seem that the phenomena might have more in common than had been apparent. In the Newtonian description, there is only one cause for a change in motion, which we call force. Forces may be of different types, but they all produce changes in motion according to the same rules.

Any acceleration that can be produced by a magnetic force can equally well be produced by an appropriately controlled stream of water.

We can speak of two forces as being equal if they produce the same change in motion when applied in the same situation, which means that they pushed or pulled equally hard in the same direction. To recapitulate briefly, a force is when a pair of objects push or pull on each other, and one newton is the force required to accelerate a 1-kg object from rest to a speed of 1 m/s in 1 second.

MANY FORCES ON OBJECT

As if we hadn't kicked poor Aristotle around sufficiently, his theory has another important flaw, which is important to discuss because it corresponds to an extremely common student misconception. Aristotle conceived of forced motion as a relationship in which one object was the boss and the other "followed orders." It therefore would only make sense for an object to experience one force at a time, because an object couldn't follow orders from two sources at once.

In the Newtonian theory, forces are numbers, not orders, and if more than one force acts on an object at once, the result is found by adding up all the forces. It is unfortunate that the use of the English word "force" has become standard, because to many people it suggests that you are "forcing" an object to do something. The force of the earth's gravity cannot "force" a boat to sink, because there are other forces acting on the boat. Adding them up gives a total of zero, so the boat accelerates neither up nor down.

EXERTING FORCES ON EACH OTHER AT DISTANCE.

Aristotle declared that forces could only act between objects that were touching, probably because he wished to avoid the type of occult speculation that attributed physical phenomena to the influence of a distant and invisible pantheon of gods. He was wrong, however, as you can observe when a magnet leaps onto your refrigerator or when the planet earth exerts gravitational forces on objects that are in the air. Some types of forces, such as friction, only operate between objects in contact, and are called contact forces. Magnetism, on the other hand, is an example of a noncontact force. Although the magnetic force gets stronger when the magnet is closer to your refrigerator, touching is not required.

WEIGHT

In physics, an object's weight, F_W, is defined as the earth's gravitational force on it. The SI unit of weight is therefore the Newton. People commonly refer to the kilogram as a unit of weight, but the kilogram is a unit of mass, not weight. Note that an object's weight is not a fixed property of that object. Objects weigh more in some places than in others, depending on the local strength of gravity. It is their mass that always stays the same. A baseball pitcher who can throw a 90-mile-per-hour fastball on earth would not be able to throw any faster on the moon, because the ball's inertia would still be the same.

POSITIVE AND NEGATIVE SIGNS OF FORCE

We'll start by considering only cases of one-dimensional centre-of-mass motion in which all the forces are parallel to the direction of motion, i.e., either directly forward or backward. In one dimension, plus and minus signs can be used to indicate directions of forces, as shown in figure c. We can then refer generically to addition of forces, rather than having to speak sometimes of addition and sometimes of subtraction. We add the forces shown in the figure and get 11 N. In general, we should choose a one-dimensional coordinate system with its x axis parallel the direction of motion. Forces that point along the positive x axis are positive, and forces in the opposite direction are negative. Forces that are not directly along the x axis cannot be immediately incorporated into this scheme, but that's OK, because we're avoiding those cases for now.

EQUILIBRIUM

When all the forces which act upon an object are balanced, then the object is said to be in a state of equilibrium. The forces are considered to bebalanced if the rightward forces are balanced by the leftward forces and the upward forces are balanced by the downward forces. This however does not necessarily mean that all the forces are*equal* to each other. Consider the two objects pictured in the force diagram shown below. Note that the two objects are at equilibrium because the forces which act upon them are balanced; however, the individual forces are not equal to each other. The 50 N force is not equal to the 30 N force.

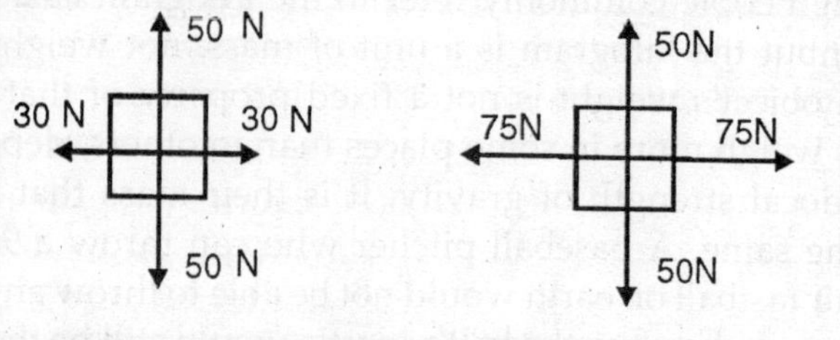

These two objects are at equilibrium since the forces are balanced. However, the forces are not equal.

If an object is at equilibrium, then the forces are balanced. *Balanced* is the key word which is used to describe equilibrium situations. Thus, the net force is zero and the acceleration is 0 m/s/s. Objects at equilibrium must have an acceleration of 0 m/s/s. This extends from Newton's first law of motion. But having an acceleration of 0 m/s/s does not mean the object is at rest.

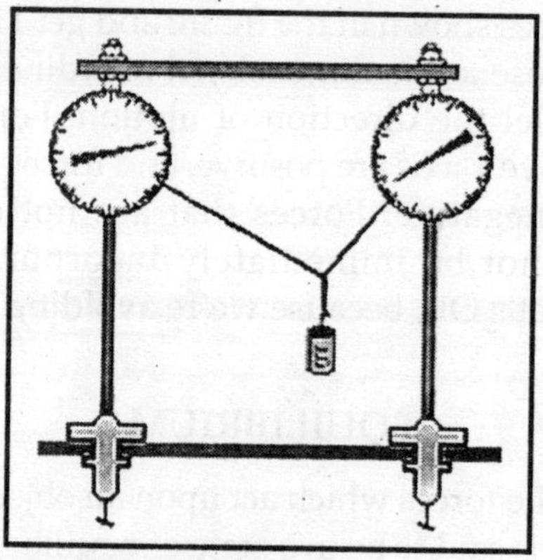

An object at equilibrium is either:

- At rest and staying at rest, or
- In motion and continuing in motion with the same speed and direction.
- This too extends from Newton's first law of motion.

If an object is at rest and is in a state of equilibrium, then we would say that the object is at "static equilibrium." "Static" means *stationary* or *at rest.*

A common physics lab is to hand an object by two or more strings and to measure the forces which are exerted at angles upon the object to support its weight. The state of the object is analysed in terms of the forces acting upon the object.

The object is a *point*on a string upon which three forces were acting. If the object is at equilibrium, then the net force acting upon the object should be 0 Newtons.

Thus, if all the forces are added together as vectors, then the resultant force (the vector sum) should be 0 Newtons. (Recall that the net force is "the vector sum of all the forces" or the resultant of adding all the individual forces head-to-tail.) Thus, an accurately drawn vector addition diagram can be constructed to determine the resultant. Sample data for such a lab are shown below.

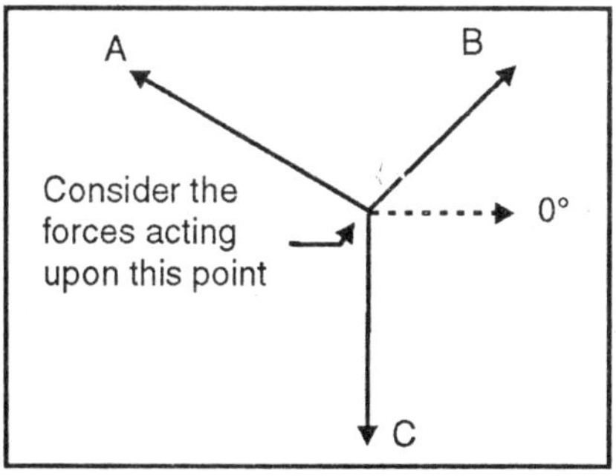

	Force A	Force B	Force C
Magnitude	3.4 N	9.2N	9.8 N
Direction	161 deg.	70 deg.	270 deg

For most students, the resultant was 0 Newtons (or at least very close to 0 N). This is what we expected - since the object was at equilibrium, the net force (vector sum of all the forces) should be 0 N.

Another way of determining the net force (vector sum of all the forces) involves using the trigonometric functions to resolve each force into its horizontal and vertical components. Once the components are known, they can be compared to see if the vertical forces are balanced and if the horizontal forces are balanced.

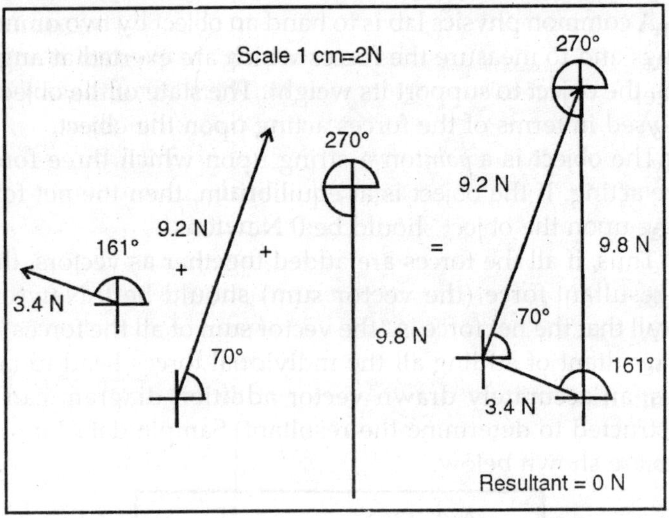

The diagram below shows vectors A, B, and C and their respective components. For vectors A and B, the vertical components can be determined using the sine of the angle and the horizontal components can be analysed using the cosine of the angle. The magnitude and direction of each component for the sample data are shown in the table below the diagram.

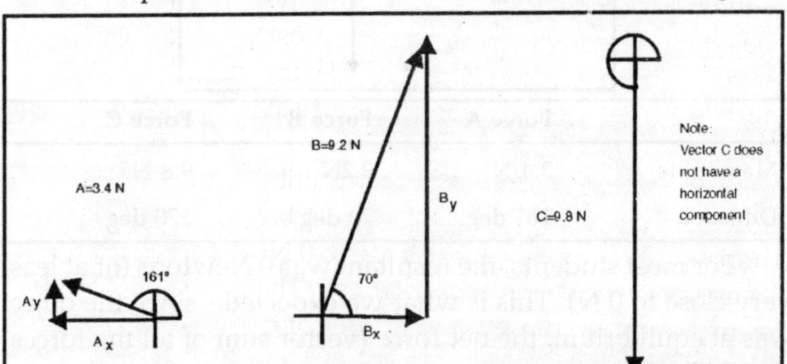

The data in the table above show that the forces *nearly* balance. An analysis of the horizontal components show that the leftward component of A *nearly* balances the rightward component of B. An analysis of the vertical components show that the sum of the upward components of A + B *nearly* balance the downward component of C. The vector sum of all the forces is (*nearly*) equal to 0 Newtons. But what about the 0.1 N difference between rightward and leftward forces and the 0.2 N difference between the upward and downward forces? Why do the components of force only *nearly* balance?

The sample data used in this analysis are the result of measured data from an actual experimental setup. The difference between the actual results and the expected results is due to the error incurred when measuring force A and force B. We would have to conclude that this low margin of experimental error reflects an experiment with excellent results. We could say it's "close enough for government work."

The above analysis of the forces acting upon an object in equilibrium is commonly used to analyse situations involving objects at static equilibrium. The most common application involves the analysis of the forces acting upon a sign that is at rest. For example, consider the picture at the right which hangs on a wall.

The picture is in a state of equilibrium, and thus all the forces acting upon the picture must be balanced. That is, all horizontal components must add to 0 Newtons and all vertical components must add to 0 Newtons. The leftward pull of cable A must balance the rightward pull of cable B and the sum of the upward pull of cable A and cable B must balance the weight of the sign.

Suppose the tension in both of the cables is measured to be 50 N and that the angle which each cable makes with the horizontal is known to be 30 degrees. What is the weight of the sign? This question can be answered by conducting a force analysis using trigonometric functions. The weight of the sign is equal to the sum of the upward components of the tension in the two cables. Thus, a trigonometric function can be used to determine this vertical component. A diagram and accompanying work is shown below.

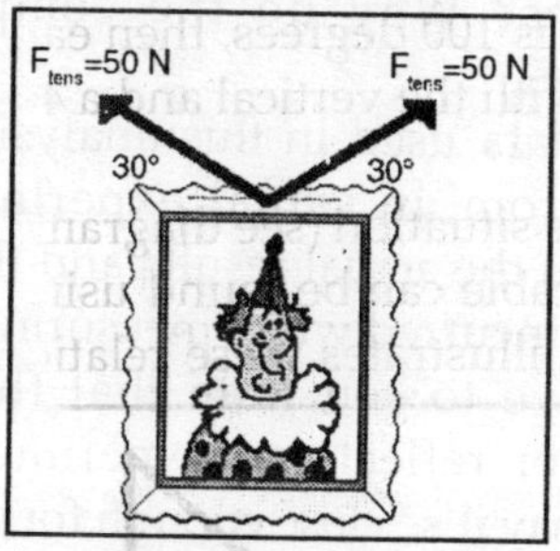

Since each cable pulls upwards with a force of 25 N, the total upward pull of the sign is 50 N. Therefore, the force of gravity (also known as weigh t) is 50 N, down. The sign weighs 50 N. In the above problem, the tension in the cable and the angle which the cable makes with the horizontal are used to determine the weight of the sign. The idea is that the tension, the angle, and the weight are related. If the any two of these three are known, then the third quantity can be determined using trigonometric functions. As another example which illustrates this idea, consider the symmetrical hanging of a sign as shown:

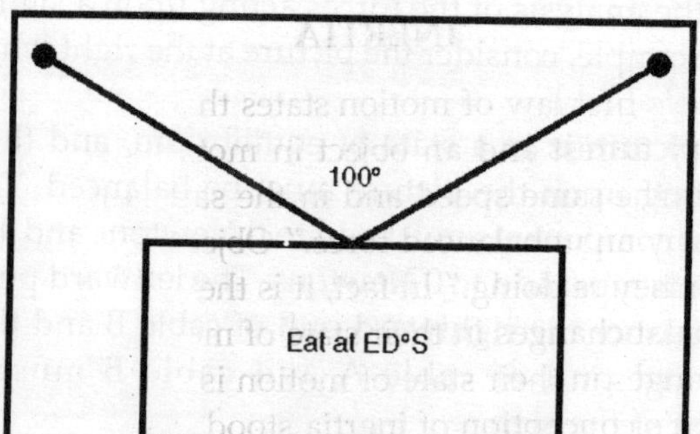

If the sign is known to have a mass of 5 kg and if the angle between the two cables is 100 degrees, then the tension in the cable can be determined. Assuming that the sign is at equilibrium (a good assumption if it is remaining at rest), the two cables must supply enough upward force to balance the downward force of gravity. The force of gravity (also known as weight) is 49 N (Fgrav = m*g), so each of the two cables must pull upwards with 24.5 N of force. Since the angle between the cables is 100 degrees, then each cable must make a 50 degree angle with the vertical and a 40 degree angle with the horizontal.

A sketch of this situation (see diagram below) reveals that the tension in the cable can be found using the sine function. The triangle below illustrates these relationships.

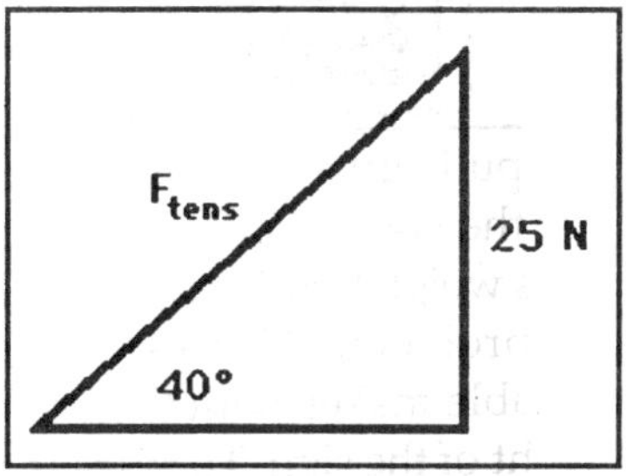

There is an important principle which emanates from some of the trigonometric calculations performed above. The principle is that as the angle with the horizontal increases, the amount of tensional force required to hold the sign at equilibrium decreases.

INERTIA

Newton's first law of motion states that "An object at rest tends to stay at rest and an object in motion tends to stay in motion with the same speed and in the same direction unless acted upon by an unbalanced force." Objects tend to "keep on doing what they're doing." In fact, it is the natural tendency of objects to resist changes in their state of motion. This tendency to resist changes in their state of motion is described as inertia

Newton's conception of inertia stood in direct opposition to more popular conceptions about motion. The dominant thought prior to Newton's day was that it was the natural tendency of objects to come to a rest position. Moving objects, so it was believed, would eventually stop moving; a force was necessary to keep an object moving. But if left to itself, a moving object would eventually come to rest and an object at rest would stay at rest; thus, the idea which dominated people's thinking for nearly 2000 years prior to Newton was that it was the natural tendency of all objects to assume a rest position.

CONCEPT OF INERTIA

Galileo, a premier scientist in the seventeenth century, developed the concept of inertia. Galileo reasoned that moving objects eventually stop because of a force called friction. In experiments using a pair of inclined planes facing each other, Galileo observed that a ball will roll down one plane and up the opposite plane to approximately the same height. If smoother planes were used, the ball would roll up the opposite plane even closer to the original height. Galileo reasoned that any difference between initial and final heights was due to the presence of friction. Galileo postulated that if friction could be entirely eliminated, then the ball would reach exactly the same height.

Galileo further observed that regardless of the angle at which the planes were oriented, the final height was almost always equal to the initial height. If the slope of the opposite incline was reduced, then the ball would roll a further distance in order to reach that original height.

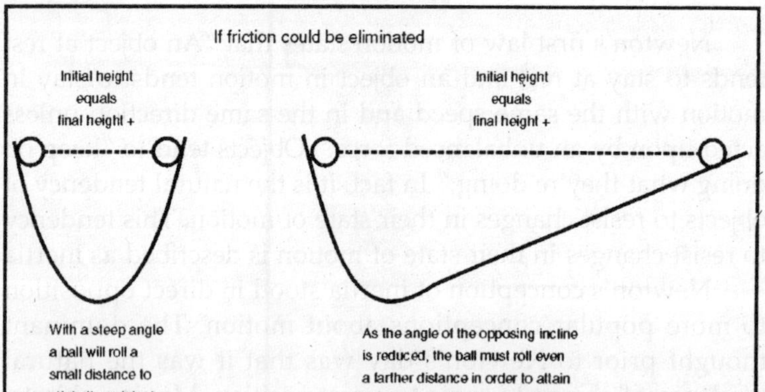

Galileo's reasoning continued - if the opposite incline was elevated at nearly a 0-degree angle, then the ball would roll almost forever in an effort to reach the original height. And if the opposing incline was not even inclined at all (that is, if it were oriented along the horizontal), then... an object in motion would continue in motion....

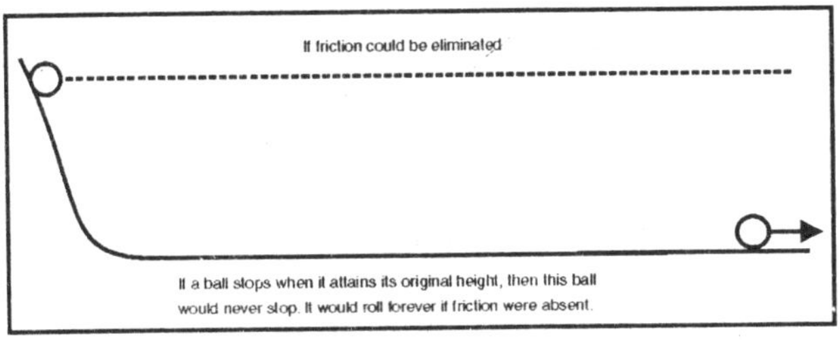

MASS AS A MEASURE OF AMOUNT OF INERTIA

All objects resist changes in their state of motion. All objects have this tendency - they have inertia. But do some objects have more of a tendency to resist changes than others? Absolutely yes! The tendency of an object to resist changes in its state of motion varies with mass. Mass is that quantity which is solelydependent upon the inertia of an object. The more inertia which an object has, the more mass it has. A more massive object has a greater tendency to resist changes in its state of motion.

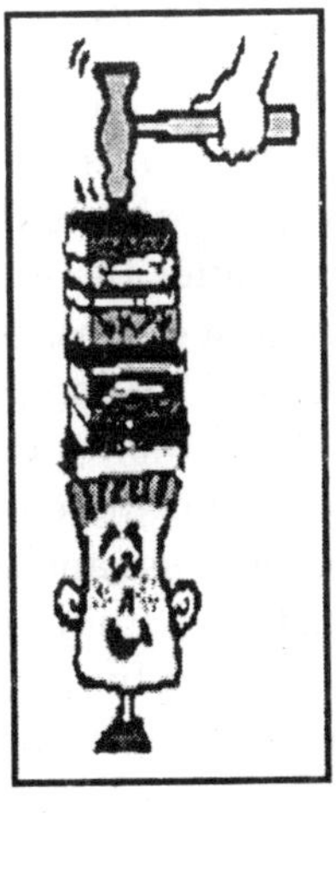

Suppose that there are two seemingly identical bricks at rest on the physics lecture table. Yet one brick consists of mortar and the other brick consists of Styrofoam. Without lifting the bricks, how could you tell which brick was the*Styrofoam brick*? You could give the bricks an identical push in an effort to change their state of motion. The brick which offers the least resistance is the brick with the least inertia - and therefore the brick with the least mass (i.e., the *Styrofoam brick*).

A common physics demonstration relies on this principle that the more massive the object, the more that object tends to resist changes in its state of motion. The demonstration goes as follows: several massive books are placed upon a teacher's head. A wooden board is placed on top of the books and a hammer is used to drive a nail into the board. Due to the large mass of the books, the force of the hammer is sufficiently resisted (inertia).

This is demonstrated by the fact that the hammer blow is not felt by the teacher. (Of course, this may explain many of the observations which you previously have made concerning.") A common variation of this demonstration involves braking a brick over the teacher's hand using the swift blow of a hammer. The massive bricks resist the force and the hand is not hurt.

MASS

We use the word mass to talk about how much matter there is in something. (Matter is anything you can touch physically.) On Earth, we weigh things to figure out how much mass there is. The more matter there is, the more something will weigh. Often, the amount of mass something has is related to its size, but not always. A balloon blown up bigger than your head will still have less matter inside it than your head (for most people, anyhow) and therefore less mass.

The difference between mass and weight is that weight is determined by how much something is pulled by gravity. If we are comparing two different things to each other on Earth, they are pulled the same by gravity and so the one with more mass weighs more. But in space, where the pull of gravity is very small, something can have almost no weight. It still has matter in it, though, so it still has mass.

RIGID BODY

A rigid body is a system of particles in which the distances between the particles do not vary. To describe the motion of a rigid body we use two systems of coordinates, a space-fixed system X, Y, Z, and a moving system x, y, z, which is rigidly fixed in the body and participates in its motion.

Let the origin of the body-fixed system be the body's centre of mass (CM). The orientation of the axes of that system relative to the axes of the space-fixed system is given by three independent angles. The vector R points from the origin of the spaced-fixed system to the CM of the body. Thus a rigid body is a mechanical system with six degrees of freedom.

Let r denote the position of an arbitrary point P in the body-fixed system. In the space fixed system its position is given by r + R, and its velocity is,

$$v = d(R + r)/dt = dR/dt + dr/dt = V + W \times r.$$

Here V is the velocity of the CM and Ω is the angular velocity of the body. The direction of Ω is along the axis of rotation and $\Omega = d\phi/dt$.

The kinetic energy of the body is:

$$T = (1/2)Sm_i v_i^2 = (1/2)Sm_i(V + W \ldots\ldots \times r)^2.$$

We rewrite,

$$T = (1/2)MV^2 + (1/2)Sm_i(W^2 r_i^2 - (W \times r_i)^2),$$

$$M = Sm_i, \quad Sm_i r_i = 0.$$

We find $T = T_{CM} + T_{rot}$, i.e. the kinetic energy is the sum of the kinetic energy of the motion of the CM and the kinetic energy of the rotation about the CM. In component form we write

$$T_{rot} = \frac{1}{2}\sum_{ij}\Omega_i\Omega_i\sum_k m_k\left(\sum_i (x_k)_i^2\,\delta_{ij} - (x_k)_i\,(x_k)_j\right) = \frac{1}{2}\sum_{ij} 1_{ij}\Omega_i\Omega_j$$

Here,

$$1_{ij} = \sum_k m_k\left(\sum_l (x_k)_i^2\,\delta_{ij} - (x_k)_i\,(x_k)_j\right)$$

is the inertia tensor. The W_i are the components of W along the axis of the body fixed system. For a continuous system.

$$\sum_k m_{k\rightarrow} \int dm = \int \rho dV$$

By appropriate choice of the orientation of the body-fixed coordinate system the inertia tensor can be reduced to diagonal form. The directions of the axes x_i are then called the principal axes of inertia and the diagonal components of the tensor are then called the principal moments of inertia.

Then,

$$T_{rot} = \frac{1}{2}\left(I_1\Omega_1^2 + I_2\Omega_2^2 + I_3\Omega_3^2\right).$$

Definitions:

- *Asymmetrical top*: $I_1 \neq I_2, I_1 \neq I_3, I_2 \neq I_3$
- *Symmetrical top*: $I_1 = I_2 \neq I_3$
- *Spherical top*: $I_1 = I_2 = I_3$

Let L denote the angular momentum about the CM of the body.

$$L = \sum_k m_k r_k \times (\Omega \times r_k) = \sum_k m_k \left(r_k^2\Omega - r_k(r_k \cdot \Omega)\right),$$

which in component form yields,

$$L_i = \sum_j \sum_k m_k \left(\sum_l (x_k)_l^2 \delta_{ij} - (x_k)_i (x_k)_j \right) \Omega_j = \sum_j I_{ij}\Omega_j.$$

If x_1, x_2, and x_3 are the principal axes of inertia, then,

$$L_1 = I_1\Omega_1,\ L_2 = I_2\Omega_2,\ L_3 = I_3\Omega_3.$$

The Lagrangian of a rigid body is,

$$L_i = \sum_j \sum_k m_k \left(\sum_l (x_k)_l^2 \delta_{ij} - (x_k)_i (x_k)_j \right) \Omega_j = \sum_j I_{ij}\Omega_j.$$

EQUATIONS OF MOTION

The equations of motion of a rigid body are dP/dt = F, where P = total momentum and F = external forces, and dL/dt = τ, where L = angular momentum about CM and τ = total torque produced by external forces.

Let A be an arbitrary vector and dA/dt its rate of change

with respect to the space fixed axes, d'A/dt its rate of change with respect to the body fixed axes.

$$dA/dt = d'A/dt + \Omega \times A.$$

Therefore, $dP/dt = d'P/dt + \Omega \times P = F$ and $dL/dt = d'L/dt + \Omega . \times L = \tau$. Let the body fixed axes be the principal axes of inertia of the body.

Then, $L_i = I_i\Omega_i$ and $d'L_i/dt = I_i d\Omega_i/dt$, and we have Euler's equations:

$$I_1 \frac{d\Omega_1}{dt} + (I_3 - I_2)\Omega_2\Omega_3 = \tau_1$$

$$I_2 \frac{d\Omega_2}{dt} + (I_1 - I_3)\Omega_3\Omega_1 = \tau_2$$

$$I_3 \frac{d\Omega_3}{dt} + (I_2 - I_1)\Omega_1\Omega_2 = \tau_3$$

FORMULAS

Centre of mass:

$$x_{CM} = \frac{\sum m_i x_i}{M}, y_{CM} = \frac{\sum m_i y_i}{M}, z_{CM} = \frac{\sum m_i z_i}{M}$$

$$Mv_{CM} = \sum m_i v_i = \sum p_i = p_{tot}$$

$$Ma_{CM} = \sum m_i a_i = \sum F_i = F_{tot}$$

NEWTON'S LAWS OF MOTION

NEWTON'S FIRST LAW

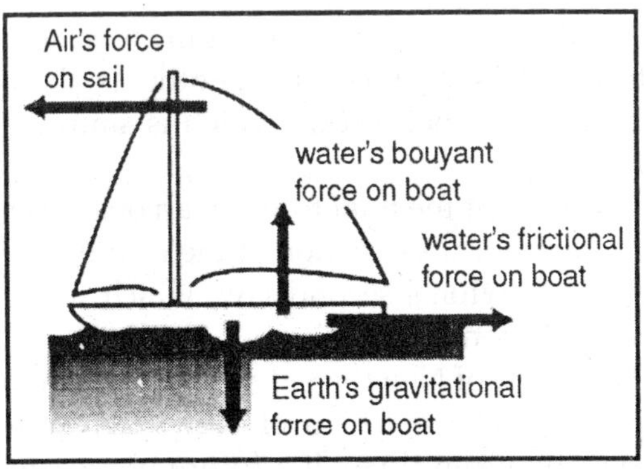

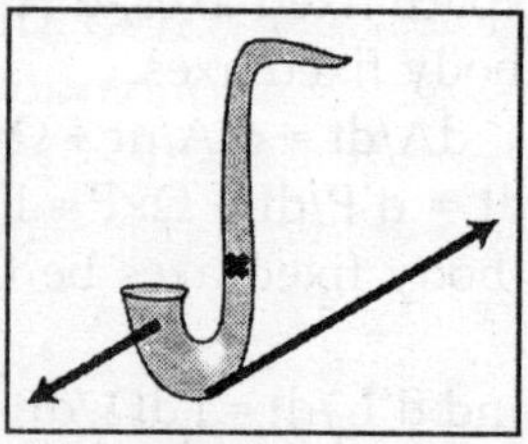

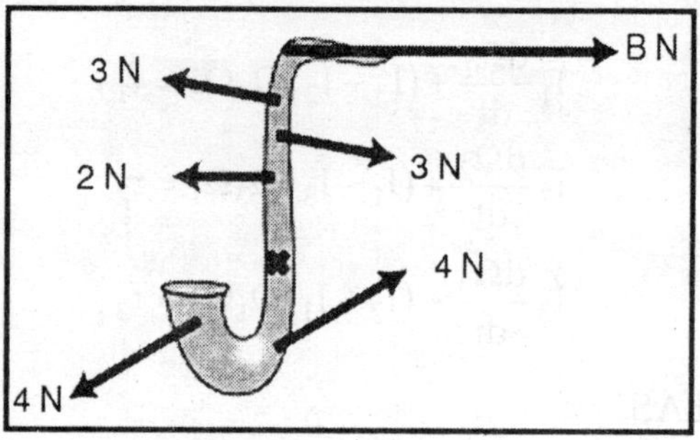

We are now prepared to make a more powerful restatement of the principle of inertia.

NEWTON'S FIRST LAW

If the total force on an object is zero, its centre of mass continues in the same state of motion.

In other words, an object initially at rest is predicted to remain at rest if the total force on it is zero, and an object in motion remains in motion with the same velocity in the same direction. The converse of Newton's first law is also true: if we observe an object moving with constant velocity along a straight line, then the total force on it must be zero.

In a future physics course or in another textbook, you may encounter the term "net force," which is simply a synonym for total force.

What happens if the total force on an object is not zero? It accelerates. Numerical prediction of the resulting acceleration is the topic of Newton's second law, which we'll discuss in the following section.

This is the first of Newton's three laws of motion. It is not important to memorize which of Newton's three laws are numbers one, two, and three. If a future physics teacher asks you something like, "Which of Newton's laws are you thinking of," a perfectly acceptable answer is "The one about constant velocity when there's zero total force." The concepts are more important than any specific formulation of them. Newton wrote in Latin, and I am not aware of any modern textbook that uses a verbatim translation of his statement of the laws of motion.

Clear writing was not in vogue in Newton's day, and he formulated his three laws in terms of a concept now called momentum, only later relating it to the concept of force. Nearly all modern texts, including this one, start with force and do momentum later.

Example: An Elevator

- An elevator has a weight of 5000 N. Compare the forces that the cable must exert to raise it at constant velocity, lower it at constant velocity, and just keep it hanging.
- In all three cases the cable must pull up with a force of exactly 5000 N. Most people think you'd need at least a little more than 5000 N to make it go up, and a little less than 5000 N to let it down, but that's incorrect. Extra force from the cable is only necessary for speeding the car up when it starts going up or slowing it down when it finishes going down. Decreased force is needed to speed the car up when it gets going down and to slow it down when it finishes going up. But when the elevator is cruising at constant velocity, Newton's first law says that you just need to cancel the force of the earth's gravity.

To many students, the statement in the example that the cable's upward force "cancels" the earth's downward gravitational force implies that there has been a contest, and the cable's force has won, vanquishing the earth's gravitational force and making it disappear.

That is incorrect. Both forces continue to exist, but because they add up numerically to zero, the elevator has no centre-of-mass acceleration. We know that both forces continue to exist because they both have side-effects other than their effects on the car's centre-of-mass motion.

The force acting between the cable and the car continues to produce tension in the cable and keep the cable taut. The earth's gravitational force continues to keep the passengers (whom we are considering as part of the elevator-object) stuck to the floor and to produce internal stresses in the walls of the car, which must hold up the floor.

Example: Terminal Velocity for Falling Objects

- An object like a feather that is not dense or streamlined does not fall with constant acceleration, because air resistance is nonnegligible. In fact, its acceleration tapers off to nearly zero within a fraction of a second, and the feather finishes dropping at constant speed (known as its terminal velocity). Why does this happen?
- Newton's first law tells us that the total force on the feather must have been reduced to nearly zero after a short time. There are two forces acting on the feather: a downward gravitational force from the planet earth, and an upward frictional force from the air. As the feather speeds up, the air friction becomes stronger and stronger, and eventually it cancels out the earth's gravitational force, so the feather just continues with constant velocity without speeding up any more.

The situation for a skydiver is exactly analogous. It's just that the skydiver experiences perhaps a million times more gravitational force than the feather, and it is not until she is falling very fast that the force of air friction becomes as strong as the gravitational force. It takes her several seconds to reach terminal velocity, which is on the order of a hundred miles per hour.

GENERAL COMBINATIONS OF FORCES

It is too constraining to restrict our attention to cases where all the forces lie along the line of the centre of mass's motion. For one thing, we can't analyse any case of horizontal motion, since any object on earth will be subject to a vertical gravitational force! For instance, when you are driving your car down a straight road, there are both horizontal forces and vertical forces.

However, the vertical forces have no effect on the centre of mass motion, because the road's upward force simply counteracts the earth's downward gravitational force and keeps the car from sinking into the ground.

Later in the book we'll deal with the most general case of many forces acting on an object at any angles, using the mathematical technique of vector addition, but the following slight generalization of Newton's first law allows us to analyse a great many cases of interest:

Suppose that an object has two sets of forces acting on it, one set along the line of the object's initial motion and another set perpendicular to the first set. If both sets of forces cancel, then the object's centre of mass continues in the same state of motion.

Example: A Passenger Riding the Subway

- Describe the forces acting on a person standing in a subway train that is cruising at constant velocity.
- No force is necessary to keep the person moving relative to the ground. He will not be swept to the back of the train if the floor is slippery. There are two vertical forces on him, the earth's downward gravitational force and the floor's upward force, which cancel. There are no horizontal forces on him at all, so of course the total horizontal force is zero.

Example: Forces on a Sailboat

- If a sailboat is cruising at constant velocity with the wind coming from directly behind it, what must be true about the forces acting on it?
- The forces acting on the boat must be canceling each other out. The boat is not sinking or leaping into the air, so evidently the vertical forces are canceling out.

The vertical forces are the downward gravitational force exerted by the planet earth and an upward force from the water.

The air is making a forward force on the sail, and if the boat is not accelerating horizontally then the water's backward frictional force must be canceling it out.

Contrary to Aristotle, more force is not needed in order to maintain a higher speed. Zero total force is always needed to maintain constant velocity. Consider the following made-up numbers:

	Boat moving at a low, constant velocity	**Boat moving at a high, constant velocity**
Forward force of the wind on the sail ...	10,000 N	20,000 N
Backward force of the water on the hull ...	–10,000 N	–20,000 N
Total force on the boat ...	0 N	0 N

The faster boat still has zero total force on it. The forward force on it is greater, and the backward force smaller (more negative), but that's irrelevant because Newton's first law has to do with the total force, not the individual forces.

This example is quite analogous to the one about terminal velocity of falling objects, since there is a frictional force that increases with speed. After casting off from the dock and raising the sail, the boat will accelerate briefly, and then reach its terminal velocity, at which the water's frictional force has become as great as the wind's force on the sail.

Example: A Car Crash

- If you drive your car into a brick wall, what is the mysterious force that slams your face into the steering wheel?
- Your surgeon has taken physics, so she is not going to believe your claim that a mysterious force is to blame. She knows that your face was just following

Newton's first law. Immediately after your car hit the wall, the only forces acting on your head were the same canceling-out forces that had existed previously: the earth's downward gravitational force and the upward force from your neck. There were no forward or backward forces on your head, but the car did experience a backward force from the wall, so the car slowed down and your face caught up.

Discussion Questions

- Newton said that objects continue moving if no forces are acting on them, but his predecessor Aristotle said that a force was necessary to keep an object moving. Why does Aristotle's theory seem more plausible, even though we now believe it to be wrong? What insight was Aristotle missing about the reason why things seem to slow down naturally? Give an example.
- In the figure what would have to be true about the saxophone's initial motion if the forces shown were to result in continued one-dimensional motion of its centre of mass?
- This figure requires an ever further generalization of the preceding discussion. After studying the forces, what does your physical intuition tell you will happen? Can you state in words how to generalize the conditions for one-dimensional motion to include situations like this one?

Newton's Second Law Example

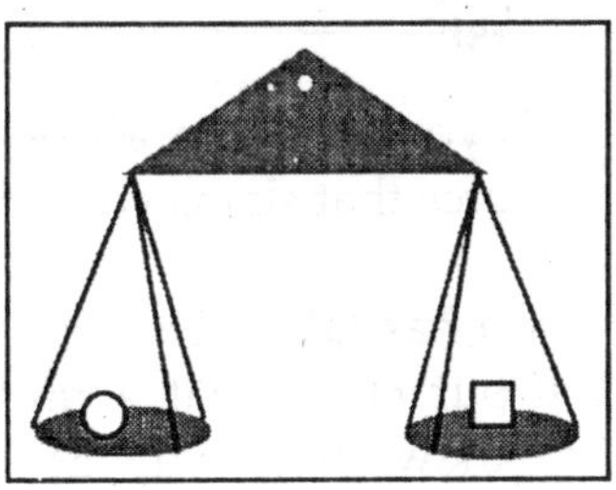

A simple double-pan balance works by comparing the weight forces exerted by the earth on the contents of the two pans. Since the two pans are at almost the same location on the earth's surface, the value of g is essentially the same for each one, and equality of weight therefore also implies equality of mass.

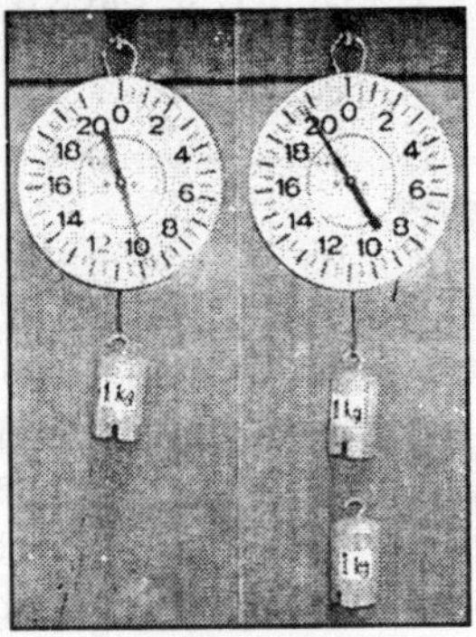

x munit	t sunit
10	1.84
20	2.86
30	3.80
40	4.67
50	5.53
60	6.38
70	7.23
80	8.10
90	8.96
100	9.83

Discussion Question

What about cases where the total force on an object is not zero, so that Newton's first law doesn't apply? The object will have an acceleration. The way we've defined positive and negative signs of force and acceleration guarantees that positive forces produce positive accelerations, and likewise for negative values. How much acceleration will it have? It will clearly depend on both the object's mass and on the amount of force.

Experiments with any particular object show that its acceleration is directly proportional to the total force applied to it. This may seem wrong, since we know of many cases where small amounts of force fail to move an object at all, and larger forces get it going. This apparent failure of proportionality actually results from forgetting that there is a frictional force in addition to the force we apply to move the object. The object's acceleration is exactly proportional to the total force on it, not to any individual force on it. In the absence of friction, even a very tiny force can slowly change the velocity of a very massive object. Experiments also show that the acceleration is inversely proportional to the object's mass, and combining these two proportionalities gives the following way of predicting the acceleration of any object:

Newton's Second Law

$$a=F_{total/m}$$

where,

m is an object's mass

F_{total} is the sum of the forces acting on it

a is the acceleration of the object's centre of mass.

We are presently restricted to the case where the forces of interest are parallel to the direction of motion.

Example: An Accelerating Bus

A VW bus with a mass of 2000 kg accelerates from 0 to 25 m/s (freeway speed) in 34 s. Assuming the acceleration is constant, what is the total force on the bus?

We solve Newton's second law for $F_{total}=ma$, and substitute $\Delta v/\Delta t$ for a, giving,

$$F_{total}=mDn/Dt$$
$$=(2000\text{ Kg})\ (25\text{ m/s}-0\text{ m/s})/(34\text{ s})$$
$$=1.5\text{ kN}$$

As with the first law, the second law can be easily generalized to include a much larger class of interesting situations: Suppose an object is being acted on by two sets of forces, one set lying along the object's initial direction of motion and another set acting along a perpendicular line. If the forces perpendicular to the initial direction of motion cancel out, then the object accelerates along its original line of motion according to $a=F_{total}/m$.

Relationship Between Mass and Weight

Mass is different from weight, but they're related. An apple's mass tells us how hard it is to change its motion. Its weight measures the strength of the gravitational attraction between the apple and the planet earth. The apple's weight is less on the moon, but its mass is the same. Astronauts assembling the International Space Station in zero gravity cannot just pitch massive modules back and forth with their bare hands; the modules are weightless, but not massless.

We have already seen the experimental evidence that when weight (the force of the earth's gravity) is the only force acting on an object, its acceleration equals the constant g, and g depends on where you are on the surface of the earth, but not on the mass of the object. Applying Newton's second law then allows us to calculate the magnitude of the gravitational force on any object in terms of its mass:

$$|F_W|=mg\,.$$

(The equation only gives the magnitude, i.e. the absolute value, of F_W, because we're defining g as a positive number, so it equals the absolute value of a falling object's acceleration.)

Example: Weight and Mass

It is shown that masses of one and two kilograms hung from a spring scale, which measures force in units of newtons.

Let's start with the single kilogram. It's not accelerating, so evidently the total force on it is zero: the spring scale's upward force on it is canceling out the earth's downward gravitational force. The spring scale tells us how much force it is being obliged to supply, but since the two forces are equal in strength, the spring scale's reading can also be interpreted as measuring the strength of the gravitational force, i.e., the weight of the one-kilogram mass.

The weight of a one-kilogram mass should be,

$$F_W = mg$$

$= (1.0 \text{ Kg})(9.8 \text{ m/s}^2) = 9.8 \text{ N}$

and that's indeed the reading on the spring scale.

Similarly for the two-kilogram mass, we have

$Fw = mg$

$=(2.0 \text{ Kg})(9.8 \text{ m/s}^2) = 19.6 \text{ N}.$

Example: Calculating terminal velocity

Experiments show that the force of air friction on a falling object such as a skydiver or a feather can be approximated fairly well with the equation $|F_{air}| = c\rho Av^2$, where c is a constant, ρ is the density of the air, A is the cross-sectional area of the object as seen from below, and $v$is the object's velocity. Predict the object's terminal velocity, i.e., the final velocity it reaches after a long time. As the object accelerates, its greater v causes the upward force of the air to increase until finally the gravitational force and the force of air friction cancel out, after which the object continues at constant velocity. We choose a coordinate system in which positive is up, so that the gravitational force is negative and the force of air friction is positive. We want to find the velocity at which,

$$F_{air} + Fw = 0, \text{ i.e.,}$$

$c\rho Av^2 - mg = 0$

Solving for v gives,

$$v_{terminal} = \sqrt{\frac{mg}{c\rho A}}$$

It is important to get into the habit of interpreting equations. This may be difficult at first, but eventually you will get used to this kind of reasoning.

- Interpret the equation $v_{terminal} = \sqrt{mg/c\rho A}$ in the case of r=0
- How would the terminal velocity of a 4-cm steel ball compare to that of a 1-cm ball?

- In addition to teasing out the *mathematical* meaning of an equation, we also have to be able to place it in its *physical* context. How generally important is this equation? (answer in the back of the PDF version of the book)

Discussion Questions

Show that the Newton can be reexpressed in terms of the three basic mks units as the combination kgÅ"m/s^2.

What is wrong with the following statements?

(1) "g is the force of gravity."

(2) "Mass is a measure of how much space something takes up."

Criticize the following incorrect statement: "If an object is at rest and the total force on it is zero, it stays at rest. There can also be cases where an object is moving and keeps on moving without having any total force on it, but that can only happen when there's no friction, like in outer space."

Table g gives laser timing data for Ben Johnson's 100 m dash at the 1987 World Championship in Rome. (His world record was later revoked because he tested positive for steroids.) How does the total force on him change over the duration of the race?

Violin teachers have to endure their beginning students' screeching. A frown appears on the woodwind teacher's face as she watches her student take a breath with an expansion of his ribcage but none in his belly. What makes physics teachers cringe is their students' verbal statements about forces.

Force is not a property of one object.

A great many of students' incorrect descriptions of forces could be cured by keeping in mind that a force is an interaction of two objects, not a property of one object.

Incorrect statement: "That magnet has a lot of force."

If the magnet is one millimetre away from a steel ball bearing, they may exert a very strong attraction on each other, but if they were a metre apart, the force would be virtually undetectable. The magnet's strength can be rated using certain electrical units (ampere-metres^2), but not in units of force.

Force is not a measure of an object's motion.

If force is not a property of a single object, then it cannot be used as a measure of the object's motion.

Incorrect statement: "The freight train rumbled down the tracks with awesome force."

Force is not a measure of motion. If the freight train collides with a stalled cement truck, then some awesome forces will occur, but if it hits a fly the force will be small.

Cenergy: There are two main approaches to understanding the motion of objects, one based on force and one on a different concept, called energy. The SI unit of energy is the Joule, but you are probably more familiar with the calorie, used for measuring food's energy, and the kilowatt-hour, the unit the electric company uses for billing you. Physics students' previous familiarity with calories and kilowatt-hours is matched by their universal unfamiliarity with measuring forces in units of Newtons, but the precise operational definitions of the energy concepts are more complex than those of the force concepts, and textbooks, including this one, almost universally place the force description of physics before the energy description. During the long period after the introduction of force and before the careful definition of energy, students are therefore vulnerable to situations in which, without realizing it, they are imputing the properties of energy to phenomena of force.

Incorrect statement: "How can my chair be making an upward force on my rear end? It has no power!"

Power is a concept related to energy, e.g., a 100-watt lightbulb uses up 100 joules per second of energy. When you sit in a chair, no energy is used up, so forces can exist between you and the chair without any need for a source of power.

Force is not stored or used up.

Because energy can be stored and used up, people think force also can be stored or used up.

Incorrect statement: "If you don't fill up your tank with gas, you'll run out of force."

Energy is what you'll run out of, not force.

Forces need not be exerted by living things or machines.

Transforming energy from one form into another usually requires some kind of living or mechanical mechanism. The concept is not applicable to forces, which are an interaction between objects, not a thing to be transferred or transformed.

Incorrect statement: "How can a wooden bench be making an upward force on my rear end? It doesn't have any springs or anything inside it."

No springs or other internal mechanisms are required. If the bench didn't make any force on you, you would obey Newton's second law and fall through it. Evidently it does make a force on you.

A force is the direct cause of a change in motion.

I can click a remote control to make my garage door change from being at rest to being in motion. My finger's force on the button, however, was not the force that acted on the door. When we speak of a force on an object in physics, we are talking about a force that acts directly. Similarly, when you pull a reluctant dog along by its leash, the leash and the dog are making forces on each other, not your hand and the dog. The dog is not even touching your hand.

PRINCIPLE OF WORK AND ENERGY

The change in the kinetic energy of an object is equal to the net work done on the object. This fact is referred to as the Work-Energy Principle and is often a very useful tool in mechanics problem solving. It is derivable from conservation of energy and the application of the relationships for work and energy, so it is not independent of thecon-servation laws.

It is in fact a specific application of conservation of energy. However, there are so many mechanical problems which are solved efficiently by applying this principle that it merits separate attention as a working principle.

For a straight-line collision, the net work done is equal to the average force of impact times the distance traveled during the impact. Average impact force x distance traveled = change in kinetic energy. The work-energy principle is a general principle which can be applied specifically to rotating objects.

For pure rotation, the net work is equal to the change in rotational kinetic energy:

$$W_{net} = \frac{1}{2}I\omega_f^2 - \frac{1}{2}I\omega_i^2$$

For a constant torque, the work can be expressed as,

$$W = \tau\theta$$

and for a net torque, Newton's 2nd law for rotation gives

$$W_{net} = \tau_{net}\theta = I\alpha\theta$$

Combining this last expression with the work-energy principle gives a useful relationship for describing rotational motion.

$$\omega_f^2 = \omega_i^2 + 2\alpha\theta.$$

If an object is rolling without slipping, then its kinetic energy can be expressed as the sum of the translational kinetic energy of its centre of mass plus the rotational kinetic energy about the centre of mass. The angular velocity is of course related to the linear velocity of the centre of mass, so the energy can be expressed in terms of either of them as the problem dictates, such as in the rolling of an object down an incline.

Note that the moment of inertia used must be the moment of inertia about the centre of mass. If it is known about some other axis, then the parallel axis theorem may be used to obtain the needed moment of inertia.

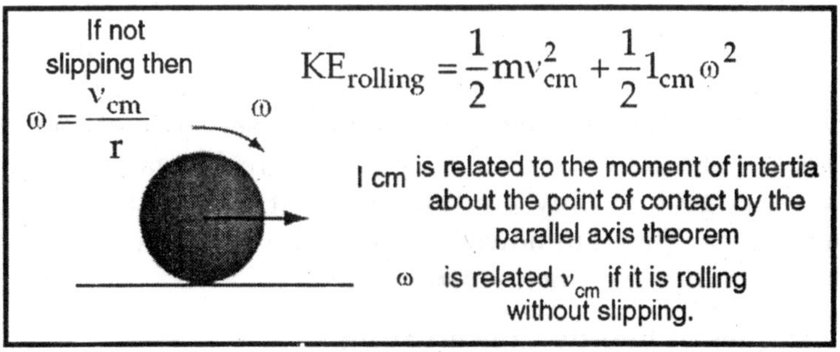

PRINCIPLE OF CONSERVATION OF ENERGY

Water is more or less conserved. So the amount of water in a reservoir can always be calculated from the amount that was there some time ago, *plus* the amount that has come in, *minus* the amount that has gone out (you may have to take account of evaporation as well as water drawn off).

Another way of saying the same thing is that water can't be made or destroyed. For there to be more, it has to come in; for there to be less it has to go out.

Energy is similar. If you take any volume of space, then the total energy inside that volume at a given time is always the amount that was there earlier, *plus* the total amount that has come in through the surface, *minus* the total amount that has gone out through the surface.

Another way of saying the same thing is that energy can't be made or destroyed. For there to be more, it must have come from somewhere; for there to be less it must have gone somewhere else.

This means that energy can quite correctly be thought of as 'rather like a fluid'. You may picture it as stored or as flowing. You may sensibly ask where it is, where it is going, where it is coming from. [It is*not exactly* like water. For example, you can measure an amount (in J) and a rate of loss or gain (in W). But you can't ask at what velocity the 'stuff' flows, because there isn't any 'stuff' whose 'particles' would have a velocity. We mention this only to keep a few niggles at bay.] Energy is 'rather like a fluid' *because* it is conserved, not the other way round. It is a calculated quantity, not observable 'stuff'. The practical teaching implication here is that it is important to do sums about energy changes – how much in, how much out – and not just to talk generally about it.

The conservation laws, such as the conservation of energy, give physics its backbone. They are not really statements of knowledge but they contain implicit assumptions and definitions. They are however tied to the natural world, and they contain experimental knowledge.

Examples of Energy Conservation

Introduce different forms of energy and discuss the interchanges. Concentrate on kinetic, gravitational potential and thermal energy. A Sankey diagram can be used to show where the energy goes when it is 'lost', such as the example shown here of energy changes per second (measured in Watts) in a motor vehicle.

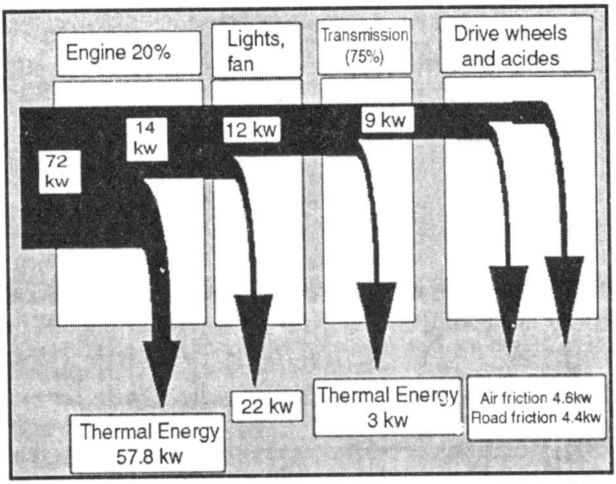

TAP 217-1: Sankey diagram for a motor vehicle. Note that some students may think that the conservation of energy is an idealized notion, and that in practical situations, energy is not conserved. This is an incorrect idea. Energy is always conserved; in practical situations, some energy may be transferred as heat energy when we do not want it to be; however, correct accounting will show that the total amount of energy is still constant.

PARALLELOGRAM LAWS

If two forces, acting at a point, are represented in magnitude and direction by the two sides of a parallelogram drawn from one of its angular points, their resultant is represented both in magnitude and direction by the diagonal of the parallelogram passing through that angular point.

RESULTANT FORCE:

If two or more forces P, Q, S,... act upon a rigid body and if a single force, R, can be found whose effect upon the body is the same as that of the forces P, Q, S, ... this single force R is called the resultant of the other forces.

Resultant of forces acting in the same direction (same straight line) is equal to their sum.

Magnitude and Direction of the Resultant of Two Forces:

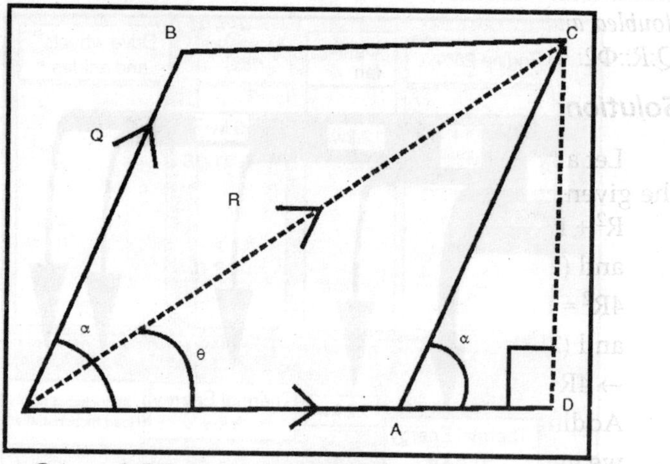

Let OA and OB represent the forces P and Q acting at a point O and inclined to each other at an angle a then the resultant R and direction 'q' (shown in figure) will be given by R = $\Phi P^2+Q^2+2PQ\cos\alpha$ and $\tan\theta = Q\sin\alpha/P+Q\cos\alpha$ Case (i) If P = Q, then tanq = tan $(\alpha/2) \Rightarrow \theta = \alpha/2$ Case

If the forces act at right angles, so that $\alpha = 90°$, we have R = ΦP^2+Q^2 and $\tan\theta = Q/P$

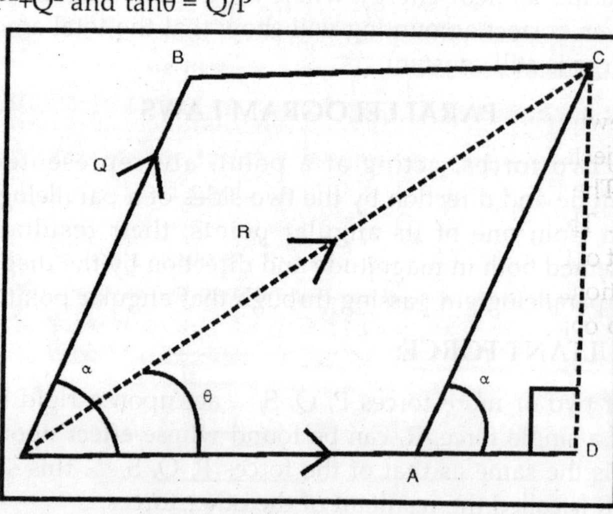

Solved Example

The resultant of two forces P and Q is R. If Q is doubled, R is doubled and when Q is reversed, R is again doubled, show that P: Q:R::Φ2: Φ3:Φ2.40

Solution:

Let a be the angle between the forces P and Q. Now from the given conditions, we have

$R^2 + P^2 + Q^2 + 2PQ \cos \alpha$ (1)

and $(2R)^2 + P^2 + (2Q)^2 + 2P(2Q)\cos \alpha$

$4R^2 = P^2 + 4Q^2 + 4PQ \cos\alpha$ (2)

and $(2R)^2 = P^2 + (-Q)^2 + 2P(-Q) \cos\alpha$

$\rightarrow 4R^2 = P^2+Q^2 - 2PQ \cos\alpha$ (3)

Adding (1) and (3),

we get $2P^2 + 2Q^2 - 5R^2 = 0$ (4)

Eliminating a from (2) and (3), we get

$P^2 + 2Q^2 - 4R^2 = 0$ (5)

From (4) and (5),

we have,

$$P^2/-8+10 = Q^2/-5+8 = R^2/4-2 \rightarrow P^2/2 = Q^2/3 = R^2/2$$

$$\rightarrow P:Q:R::\Phi 2: \Phi 3: \Phi 2$$

NEWTON'S LAW OF GRAVITATION

Newton's Universal Law of Gravitation states that any two objects exert a gravitational force of attraction on each other. The direction of the force is along the line joing the objects. The magnitude of the force is proportional to the product of the gravitational masses of the objects, and inversely proportional to the square of the distance between them. For the two objects in Figure.

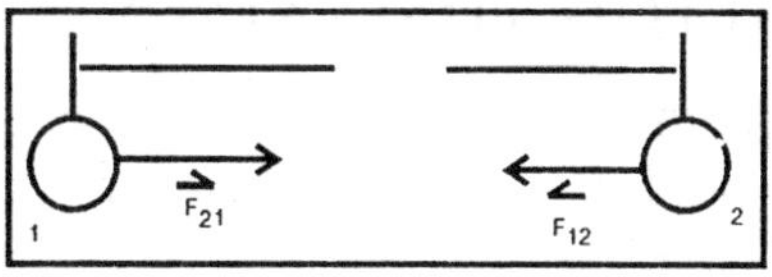

Fig. Gravitational Force Between Two Masses

- m_1 exerts a force $\bar{F}_{12}$ on m_2.
- m_2 exerts a force $\bar{F}_{21}$ on m_1.
- By Newton's third law:

$$\bar{F}_{12} = -\bar{F}_{21}$$

- The magnitude of the gravitational force is:

$$F_{12} = G\frac{m_1 m_2}{r^2}$$

- G is Newton's constant:
  $G = 6.67 \times 10^{-11}\ N\,m^2/kg^2$. (23)

The inertial mass of an object determines the amount of force needed to produce a given acceleration of that object. The gravitational mass determines the force of gravitational attraction between two bodies. In Newtonian mechanics, these two masses have no obvious connection with each other. Nonetheless, it was observed empirically that they are numerically equal. This remarkable fact was known for centuries, but remained unexplained until Einstein's General Theory of relativity.

- Newton's gravitational constant is *extremely small* when expressed in terms of laboratory sized objects: the gravitational force between two 1 *kg* objects separated by 1 *m* is only 6.67×10^{-11} Newtons.
- For an object of mass *m* near the Earth's surface:

$$F_{grav} = -G\frac{M_E}{R_E^2}m = -mg$$

- Where $M_E = 5.98 \times 10^{24}$ kg is the mass of the Earth and $R_E = 6.38 \times 10^6$ m is the radius of the earth and

$$g \equiv G\frac{M_E}{R_E^2} = 9.8\,m/s^2$$

- In agreement with the expression in Chapter 3.

GRAVITIONAL POTENTIAL ENERGY

Due to the gravitational force of attraction, any two objects with masses m_1 and m_2 located a distance r apart have the ability to do work. Hence they have potential energy. The gravitational potential energy of such objects is:

$$PE_{gray} = -G\frac{m_1m_2}{r}$$

Note:

- Recall that only differences in potential energy are physically relevant. In the above, the zero of gravitational potential energy has been arbitrarily chosen to be zero at $r = \infty$. i.e. when the objects are infinitely far apart.
- The negative sign is a consequence of the attractive nature of the gravitational force. When the objects are far apart, the gravitational force naturally moves them closer, decreasing their potential energy (i.e. making it more negative).
- Gravitational potential energy near Earth's surface: For an object of mass m a distance h above the earth's surface:

$$PE_{gray} = -G\frac{mM_E}{(R_E + h)}$$

If $h << R_E$ *we can approximate*:

$$\frac{1}{(R_E + h)} \approx \frac{1}{R_E}\left(1 - \frac{h}{R_E}\right)$$

so that:

$$PE_{gray} \approx \frac{mM_E}{R_E}\left(1 - \frac{h}{R_E}\right)$$

$$= -G\frac{mM_E}{R_E} + m\frac{GM_E}{R_E^2}$$

$$= \text{Constant} + mgh$$

Chapter 2

Force and Force Systems

FORCE

Strength or energy of body or mind; active power; vigour; might; often, an unusual degree of strength or energy; capacity of exercising an influence or producing an effect; especially, power to persuade, or convince, or impose obligation; pertinency; validity; special signification; as, the force of an appeal, an argument, a contract, or a term.

FORCE SYSTEM

The motion under balanced force system is the subject matter of Newton's first law of motion. The condition of balanced force system simply implies that the net external force on the body is zero. This is exactly the situation for which Newton's first law predicts that the body maintains its state of translational motion.

Mathematically, the net external force is zero for a balanced force system:

$$SF=0$$

This implies that linear acceleration of the point mass is zero,

$$a=0$$

For three dimensional rigid body, linear acceleration of the centre of mass is zero (we shall discuss this aspect in detail subsequently in the course),

$$a_C=0$$

In component form,

$$\Sigma F_x=ma_x=0$$

$$\Sigma F_y = ma_y = 0$$
$$\Sigma F_z = ma_z = 0$$

It is to be noted here that the condition of zero net force does not mean necessarily "rest" as may be inferred from most of the examples in the immediate surrounding. A balanced force system simply indicates that velocity of the body does not change i.e. v=constant

EQUILIBRIUM AND BALANCED FORCE SYSTEM

It is clear that a balanced force system is a necessary condition for equilibrium, but it is not the sufficient condition to ensure equilibrium. However, a close look, at the body systems that we generally consider in dynamics, reveals that many of the systems eventually are restricted to translational motion. There are following three situations under which a body is restricted to have only translational motion:

- The body is an infinitesimally small entity to be approximated as a point. In this case, forces are concurrent, meeting at a common point.
- The body is a real three dimensional entity. The forces on the body are concurrent at a common point. The common point, in turn, coincides with "centre of mass" of the body.
- The body is a real three dimensional entity. The forces on the body are not concurrent at a common point. However, the body is restrained to have only translational motion. The forces are though not concurrent, but we consider them concurrent at the "centre of mass" of the body, as no rotational motion is involved.

In the nutshell, if we can treat the motion as translational motion without any angular or rotational motion involved, then a balanced force system with zero net force alone is sufficient to meet the requirement of equilibrium of the body.

EXAMPLE

A block of mass "m" is at rest on a rough incline of angle

"θ". Find the contact forces and the net contact force on the block. A block on an incline.

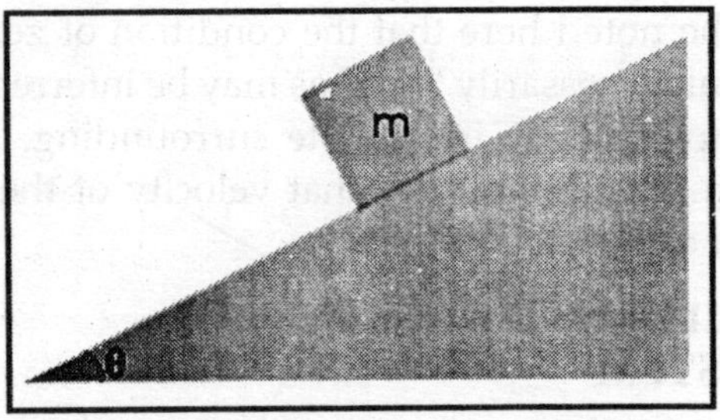

Fig. The Block Rests on the Incline.

Solution

The forces on the block are:

- Its weight
- Normal force and
- Friction force

These forces are not concurrent: Forces on the block

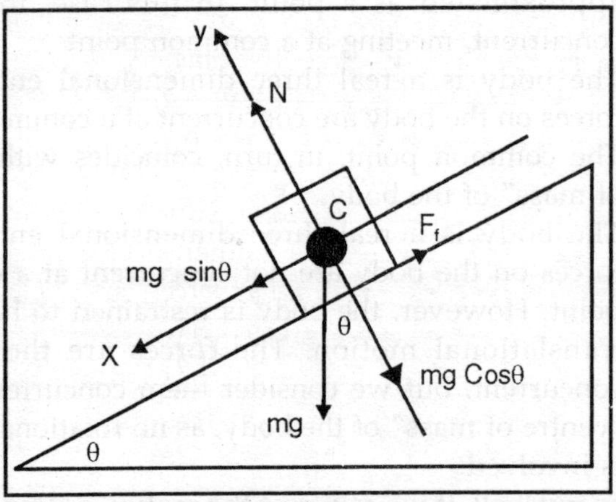

Fig. Forces are not concurrent

However, no turning effect is involved. We can, therefore, treat force system concurrent with the "centre of mass" of the block. In order to analyse the forces, we consider a coordinate system as shown in the figure. Since the block is at rest, the forces on the block are balanced:

Free Body Diagram

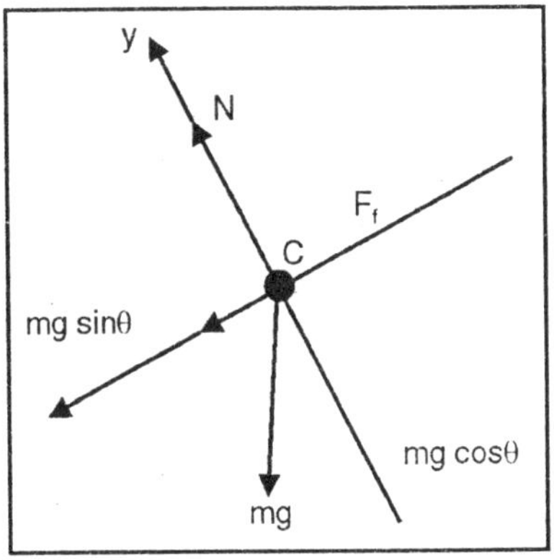

Fig. Forces are Shown with Block as a Point with Concurrent Forces

$$SF_x=mg\text{sinq}''F_f=0$$
$$SF_y=N''mg\text{cosq}=0$$

There are two contact forces (i) normal force and (ii) friction.

The friction is given by the first equation:

$$F_f=mg\text{sinq}$$

The normal force is given by the second equation:

$$N=mg\text{cosq}$$

The net contact force is vector sum of two contact forces, which are at right angle to each other.

Hence, magnitude of net contact force is:

$$F_c=\sqrt{\left(N^2+F_f^2\right)}$$
$$F_c=\sqrt{\left\{(mg)^2\left(\cos^2\theta+\sin^2\theta\right)\right\}}$$
$$F_C=mg$$

RESULTANT OF A FORCE SYSTEM

Machines and structures are usually subjected to lots of forces. When we analyse force systems, we usually are interested in the *sum* of the forces. The sum of a set of forces is known as the *resultant* of the force system. We need to know how to add forces together.

Resultant of a Set of Discrete Forces

This is straightforward. We know that forces are vectors, so they must add like vectors. We can use the usual vector machinery to sum them. The procedure is,

- Choose convenient basis vectors {i,j,k} or $\{e_1, e_2, e_2\}$
- Using geometry or trigonometry, calculate the force component along each of the three reference directions (F_x, F_y, F_z) or (F_1, F_2, F_3) You need to do this for *each* force. Let's suppose that there are *N* forces, with components,

$$\left(F_x^{(1)}, F_y^{(1)}, F_z^{(1)}\right)$$

$$\left(F_x^{(2)}, F_y^{(2)}, F_z^{(2)}\right)$$

$$\left(F_x^{(N)}, F_y^{(N)}, F_z^{(N)}\right)$$

- We can use the formula for a vector sum in terms of its components to show that the rezsultant force,

$$F_{resul\tan t} = F^{(1)} + F^{(2)} + ...F(N)$$

$$= \left(F_x^{(1)} + F_x^{(2)} + ... + F_x^{(N)}\right)i + \left(F_y^{(1)} + F_y^{(2)} + ... + F_y^{(N)}\right)j +$$

$$\left(F_y^{(1)} + F_y^{(2)} + ... + F_y^{(N)}\right)k$$

In words, the components of the resultant force vector are equal to the sum of the components of all the individual forces.

Examples

A few simple examples illustrate the idea:

1. The figure below shows a heavy box suspended from two cables. The box is subjected to a vertical gravitational force *W*, and two forces of unknown magnitude T_1, T_2 acting parallel to cables along OA, and OB, respectively. Find an expression for the vector sum of the forces (in terms of T_1 T_2), expressing your answer as components in the basis shown.

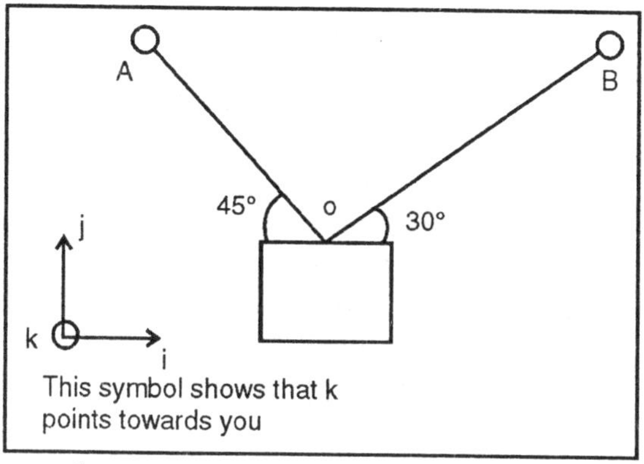

We start by expressing the forces in the basis, as outlined in the vector tutorial,

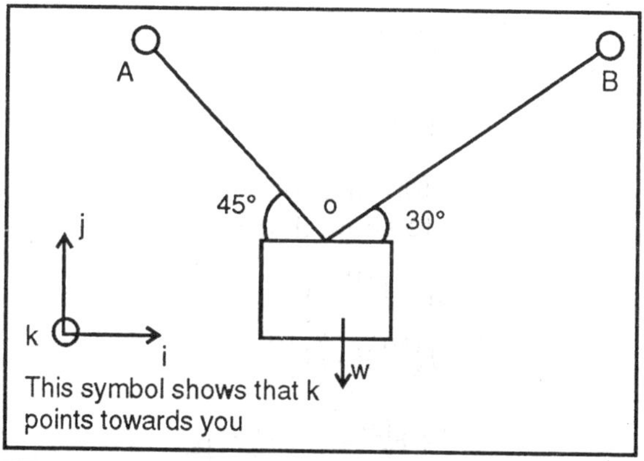

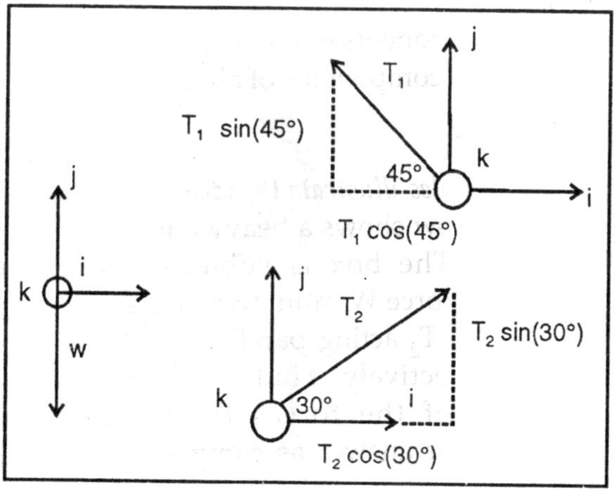

$$F_{gravity} = -Wj$$

$$F_{OA} = -T_1/\sqrt{2}i + T_1/\sqrt{2}j$$

$$F_{OB} = T_2/2i + T_2\sqrt{3}2j$$

so the sum is,

$$F_{tot} = \left(\frac{T_2\sqrt{3}}{2} - \frac{T_1}{\sqrt{2}}\right)i + \left(\frac{T_1}{\sqrt{2}} + \frac{T_2}{2} - W\right)j$$

- The sailboat shown in the picture is subjected to the following forces:
  - A force of 500N acting perpendicular to the sail
  - A weight force of 1000N acting vertically downwards
  - A buoyancy force of unknown magnitude (denoted by *B*) acting vertically upwards
  - A drag force of unknown magnitude (denoted by D) opposing forward motion of the boat
  - A drag force of unknown magnitude (denoted by *L*) opposing sideways motion of the boat

Find an expression for the resultant force.

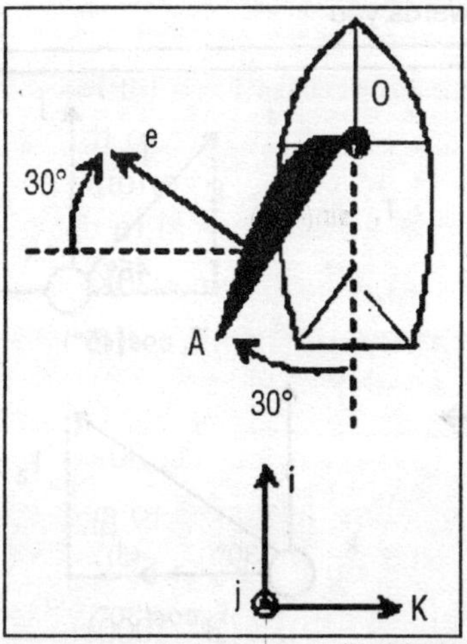

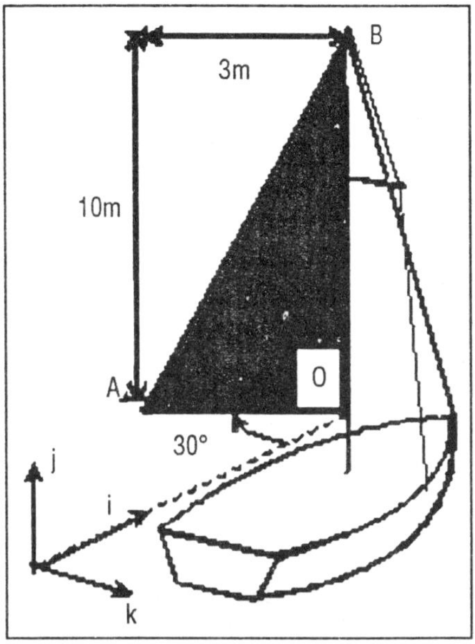

We follow the rules. The {i,j,k} vectors are shown already. Forces 2-5 are easy,

$$F^{(2)} = -1000j$$

$$F^{(3)} = Bj$$

$$F^{(4)} = -Di$$

$$F^{(5)} = LK$$

The force acting on the sail is a bit more tricky – we need to find a unit vector perpendicular to the sail. A view of the boat from above is helpful. Trigonometry shows that the components of a unit vector normal to the sail are,

$$e = \sin 30i - \cos 30K$$

(remember that a unit vector has length 1, and resolve into i and k components)

Therefore $F^{(1)} = 500(\sin 30i - \cos 30k)$

Finally we can sum the forces to get

$$F_{resultant} = F^{(1)} + F^{(2)} + F^{(3)} + F^{(4)} + F^{(5)}$$

$$= (500\sin 30 - D)i + (B - 1000)j + (L - 500\cos 30)k$$

Resultants of Distributed Loads

Many engineering problems involve *distributed* forces. Examples include pressure forces acting on aerodynamic forces or structures, hydrostatic pressure acting on submarine structures or on ships, and gravity loading.

We need to know how to calculate the resultant force exerted by continuously distributed forces (e.g. pressure distributions). We'll illustrate the general idea using 1D examples in this course, and defer more realistic 3D problems to future courses.

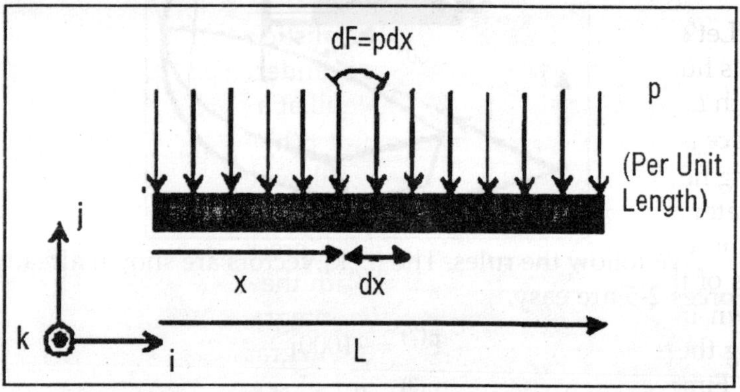

As a simple preliminary example, let's the resultant force exerted by a uniform loading acting on a beam. The beam has length *L* and is subjected to a uniformly distributed load *p* per unit length. We'll work with the coordinate system shown in the picture.

Note that the resultant force due to the pressure acting on a small piece of the beam width *dx* is,

$$dF = -pdxj$$

The total force can then be computed by summing (integrating) over the entire length of the beam,

$$F = \int_0^L -pdxj = -pLj$$

The same approach can be used to calculate resultant forces caused by more complex distributions of pressure.

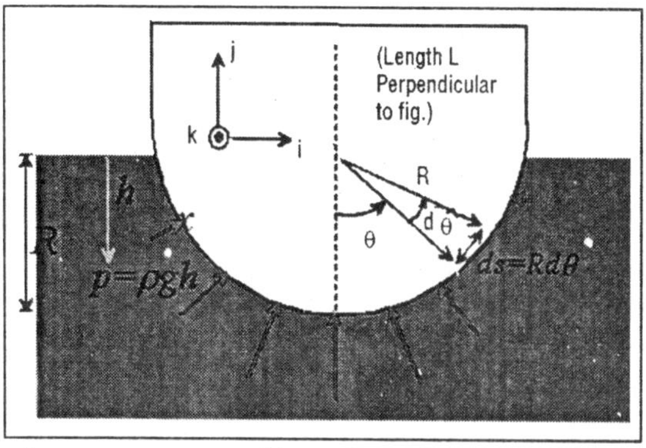

Let's do a more complicated, realistic problem. Part of a ship's hull is modeled as a circular cylinder with radius *R* and length *L*. The ship's draft is *R*. The hull at a depth *h* below the surface is subjected to a fluid pressure $p(h) = \rho gh$ per unit area acting normal to the surface of the hull, where ρ is the density of water and *g* is the acceleration of gravity. Find an expression for the resultant force exerted by the pressure acting on the sides of the hull. It's best to work with the coordinate system shown in the picture. We'll do the integral over the surface using the angle θ as the variable of integration.

First calculate the force on a small segment of the hull's surface between and The length of the segment is and so its area is The pressure acting on this segment is $\rho = \rho gh = \rho gR\cos\theta$.

The magnitude of the force acting on the segment of the hull is therefore $dF = pdA = \rho gR\cos\theta RLd\theta$

The direction of the force is towards the centre of the circle. $-\sin\theta i + \cos\theta j$ Simple geometry shows thatis a unit vector with the correct direction. Therefore the vector force acting on the segment is $dF = \rho gR\cos\theta(-\sin\theta i + \cos\theta j)RLd\theta$

The total force follows by summing (integrating) over the immersed portion of the ship's hull.

Thus,

$$F = \int_{-\pi/2}^{\pi/2} \rho gR\cos\theta(-\sin\theta i + \cos\theta j)RLd\theta$$

We can quickly do the integral using MAPLE to find that,

$$F = \frac{\pi}{2}\rho g R^2 L j$$

Of course, we could have anticipated this result using Archimedes' principle (the buoyancy force on an object is equal to the weight of water it displaces).

RESULTANT OF COPLANAR CONCURRENT FORCE SYSTEM

Coplanar concurrent forces are those which are lying in the same plane and acting at the same point. The forces can be represnted in scalar form (magnitude and angle) or in a cartesian form(components along x, y, z axes alongwith unit vectors).

Scalar Form

To find the resultant of the system of coplanar concurrent forces the following steps should be followed.

1. Resolve all the forces into their rectangular components along x-axis and y-axis.
   Note: Component of a force F along x-axis is F cos q, (where q is the angle between the force and the positive x-axis), and the component of a force F along y-axis is F sin q.
2. Take the algebraic sum of the components along x-axis and y-axis Σ Fx and Σ Fy.
3. Resultant of the force system will be calculated as
   Magnitude, R= sqrt{(Fx)(Fx)+(Fy)(Fy)}
   Direction, Angle a =Tan(–1){(Σ Fy/Σ Fx)}

Cartesian form

The forces are commonly resprented in the cartesian form with their x, y and z components as;

F=fx i + fy j + fz k

where fx, fy and fz are the components of the force along x, y and z axis and i, j, k are the unit vectors in the direction of x, y,z axis respectively, e.g., F=2 i + 3 j + 4 k

To find the resultant in the cartesian form, follow the following steps.

1. Resultant R=(Σ fx) i + (Σ fy) j + (Σ fz) k where Σ fx is the algebraic sum of the x-components of all the forces Σ fy is the algebraic sum of the y-components of all the forces Σ fz is the algebraic sum of the z-components of all the forces
2. The magnitude and direction of the resultant will be calculatd in the normal way as for any vector.

RESOLUTION AND COMPOSITION OF FORCES

The resolution of forces consists in finding two or more forces which shall have the same effect as a single force. For the weight W might be sustained by a vertical force in the direction c C equal to it; or this vertical force, it is obvious, may be resolved into two forces in the directions of the beams capable of producing the same effect as the vertical force C c.

The composition of forces consists in finding one force that shall produce the same effect as two or more forces acting in different directions. This is nothing more than the reverse of the resolution of forces, and may be accomplished in a similar manner.

If a vertical line C c be drawn through the centre of the weight, and a c be drawn parallel to the beam A C; also b c parallel to B C; then the relations between the weight and the pressures will be found by the following proportions:

- As the line C c, Is to the line C b; So is the weight W,
- To the pressure in the direction of the beam A C. Also, As the line C c,
- Is to the line C a;
- So is the weight W,
- To the pressure in the direction of the beam C B.

To those who are acquainted with the principles of mechanics the truth of the principle from which these proportions are derived requires no illustration; but such as have not had the advantage of that branch of learning may, by having recourse to the following simple experiment, not only satisfy themselves of its truth, but also render themselves more familiar with the nature of forces.* Let a thread or fine line be passed over the pulleys B and C ,

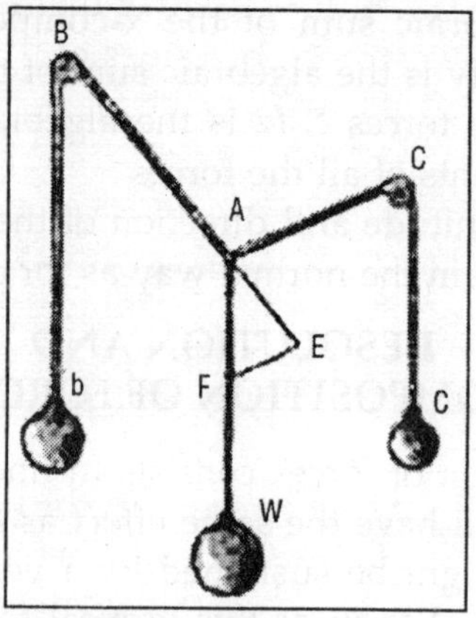

Let a known weight be attached to each end of the line, as at b and c; also let another thread be knotted to the first one at any point A, and attach a known weight to the end W. Then if the sum of the weights b and c be greater than the single weight W, there is a certain position in which the assemblage will be at rest; and if it be deranged by pulling at any of the weights it will return of itself to the same position when left at liberty.

Therefore, in that position, and in that position only, the weights will balance one another, or be in equilibrio. Now, if the positions of the threads, when the weights balance one another, were drawn upon paper; and, from a scale of equal parts, A F were made equal to the number of pounds in the weight W, and the line B A were continued to E, and the line F E drawn parallel to A C, then F E measured by the same scale of equal parts would show the number of pounds in the weight at c; also the measure of the line A E would be equal to the number of pounds in the weight b.

f the three weights he equal, then the three lines A F, F E, and A E will be equal, and the angles formed by the threads round the knot will he equal. And universally whenever the directions of three forcees are in the same plane, and meet in a point, and are in equilibria, those forces will be represented in magnitude by the three sides of a triangle drawn parallel to the directions of the forces. Consequently, if a body be kept at rest by three forces, and any two of them be represented in magnitude and direction by two sides of a triangle, the third side taken in order will represent the magnitude and direction of the other force.

Also because the sides of triangles are as the sines of the opposite angles, it follows that when three forces keep a body in equilibrio, each force is proportional to the sine of the angle made by the direction of the other two. Thus, if the weight W (Fig.) be as the sine of the angle A E F, the weight b will be as the sine of the angle A F E, etc.

It may, however, be observed, that the designs of framing are always drawn on paper to a scale; hence the proportions of the forces may be obtained immediately from the figure without the trouble of calculation, and the values so obtained will bo accurate enough for any practical purpose. This method then will be adopted in the following pages whenever ıt is found most convenient.

TURNING EFFECT OF FORCES

In physics, measure of the turning effect, or torque, produced by a force acting on a body. It is equal to the product of the force and the perpendicular distance from its line of action to the point, or pivot, about which the body will turn. The turning force around the pivot is called the moment. Its unit is the newton metre.

The moment of a force can be worked out using the formula: moment = force applied × perpendicular distance from the pivot. If the magnitude of the foıce is F newtons and the perpendicular distance is d metres then:

$$\text{moment} = Fd$$

It is easier to undo a bolt using a long spanner than a short spanner. This is because more turning force is produced at the bolt (pivot) with less effort. A long spanner is an example of a force multiplier.

In a simple balanced see-saw, the forces acting on the left- and right-hand sides of the pivot are the same. This is known as balancing moments. Moving the load at one end will cause the see-saw to become unbalanced. To regain balance, the load on the opposite side must either be increased or its position changed. This is known as the principle of moments.

The application of balancing a moment of force (or a turning force) is used by a rope walker. By holding the pole in the middle the rope walker is balancing the turning force on either side of the rope. As the rope walker moves a leg, the turning force on one side is greater and becomes unbalanced. To regain the balance, the pole is shifted to balance the turning forces on both sides of the rope.

MOMENT OF A FORCE

The Moment of a force is a measure of its tendency to cause a body to rotate about a specific point or axis. This is different from the tendency for a body to move, or translate, in the direction of the force.

In order for a moment to develop, the force must act upon the body in such a manner that the body would begin to twist. This occurs every time a force is applied so that it does not pass through the centroid of the body. A moment is due to a force not having an equal and opposite force directly along it's line of action.

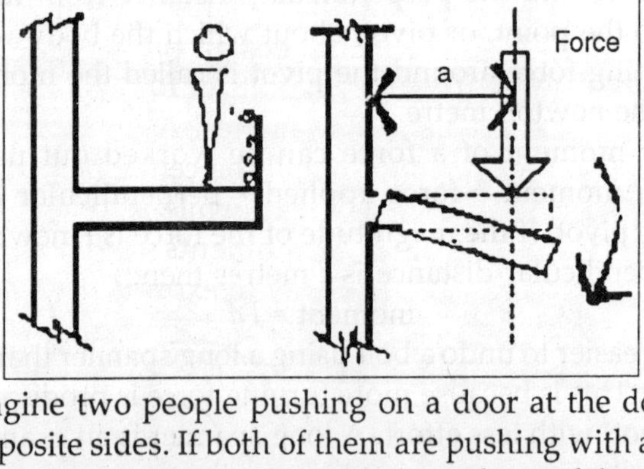

Imagine two people pushing on a door at the doorknob from opposite sides. If both of them are pushing with an equal force then there is a state of equilibrium. If one of them would suddenly jump back from the door, the push of the other person would no longer have any opposition and the door would swing away. The person who was still pushing on the door created a moment.

The magnitude of the moment of a force acting about a point or axis is directly proportinoal to the distance of the force from the point or axis. It is defined as the product of the force (F) and the moment arm (d). The moment armor lever arm is the perpendicular distance between the line of action of the force and the centre of moments.

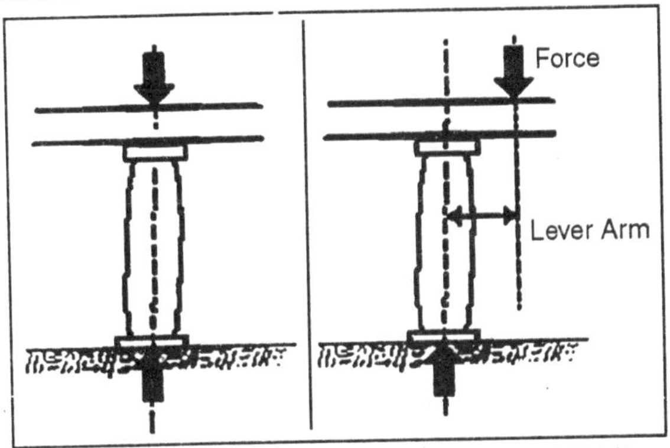

Moment = Force × Distance or M = (F)(d)

The Centre of Moments may be the actual point about which the force causes rotation. It may also be a reference point or axis about which the force may be considered as causing rotation. It does not matter as long as a specific point is always taken as the reference point. The latter case is much more common situation in structural design problems.

A moment is expressed in units of foot-pounds, kip-feet, newton-metres, or kilonewton-metres. A moment also has a sense; A clockwise rotation about the centre of moments will be considered a positive moment; while a counter-clockwise rotation about the centre of moments will be considered negative. The most common way to express a moment is,

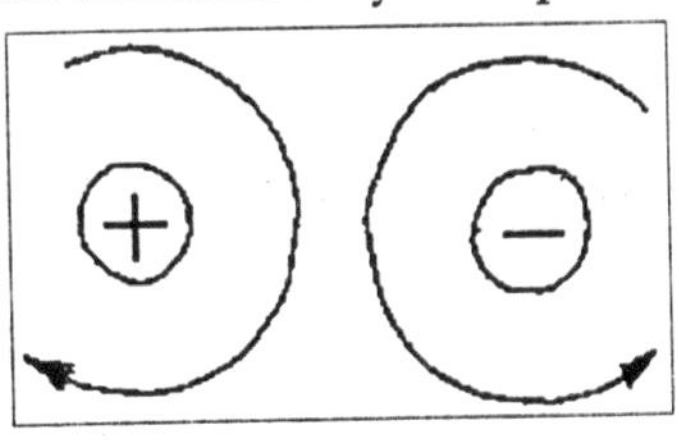

EXAMPLE

The example shows a wrench being applied to a nut. A 100 pound force is applied to it at point C, the centre of the nut. The force is applied at an x- distance of 12 inches from the nut. The centre of moments could be point C, but could also be points A or B or D.

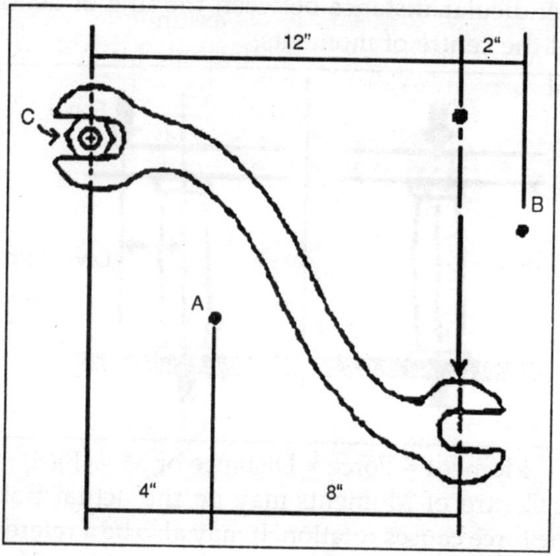

Moment about C

The moment arm for calculating the moment around point C is 12 inches. The magnitude of the moment about point C is 12 inches multiplied by the force of 100 lbs to give a total moment of 1200 inch-lbs (or 100 ft-lbs).

Moment Arm (d) = 12 inches

Magnitude (F) = 100 lbs

Moment = M = 100 lbs × 12 in. = 1200 in-lbs

Similarly, we can find the moments about any point in space.

A moment causes a rotation about a point or axis. If the moment is to be taken about a point due to a force F, then in order for a moment to develop, the line of action cannot pass through that point.

If the line of action does go through that point, the

moment is zero because the magnitude of the moment arm is zero. Such was the case for point D in the previous wrench problem. The total moment was zero because the moment arm was zero as well.

Example

As another example, let us assume that 200 pound force is applied to the wrench as indicated. The moment of the 200 pound force applied at C is zero because:

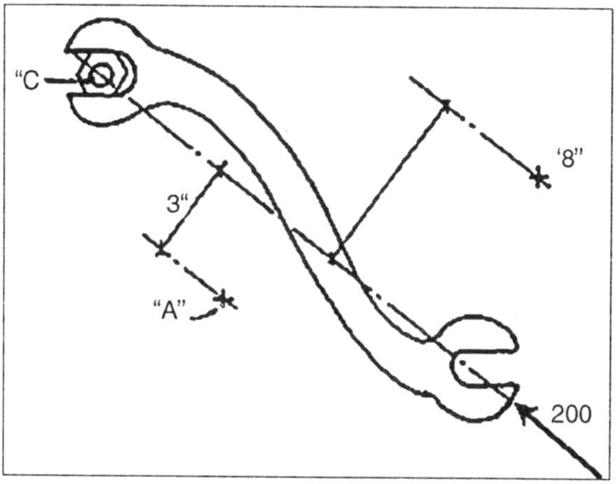

$$M = F \times d = 200 \text{ lbs} \times 0 \text{ in} = 0 \text{ in-lbs}$$

In other words, there is no tendency for the 200 pound force to cause the wrench to rotate the nut. One could increase the magnitude of the force until the bolt finally broke off (shear failure).

The moment about points X, Y, and Z would also be zero because they also lie on the line of action.

A moment can also be considered to be the result of forces detouring from a direct line drawn between the point of loading of a system and its supports. In this case, the blue force is an eccentric force.

In order for it to reach the base of the column, it must make a detour through the beam. The greater the detour, the greater the moment. The most efficient structural systems have the least amount of detours possible.

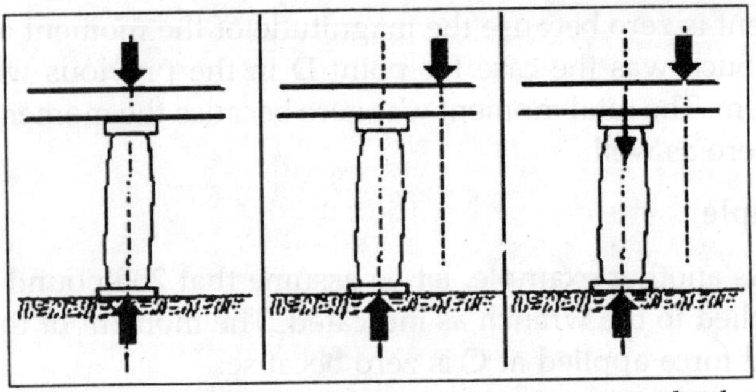

There are cases in which it is easier to calculate the moments of the componenets of a force around a certain point than it is to calculate the moment of the force itself. It could be that the determination of the perpendicular distance of the force is more difficult than determining the perpendicular distance of components of the force.

The moment of several forces about a point is simply the algebraic sum of their component moments about the same point. When adding the moments of componenets, one must take great care to be consistant with the sense of each moment. It is often prudent to note the sense next to the moment when undertaking such problems.

VARIGNON'S THEOREM

"The moment caused by the resultant force (of some system of forces) about some arbitrary point is equal to the sum of the moments due to all of the component forces of the system."

Let's break this down:

- Think about a system of forces. The simplest equivalent to this system is a single resultant force.
- Now consider some arbitrary point... any point. It doesn't even have to be physically on the system.
- As long as the line of action of the resultant of those forces in the system doesn't pass through that arbitrary point, it must cause a moment about that point.
- According to the Principle of Moments, if you calculated the moment of each force of the original

system about that same point and added them up, you would get the same moment as that calculated due to the resultant force.

So this basically tells us that if we're attempting to calculate the total moment of forces about some point, we have two ways to do it:

- Calculate each moment (from each force separately) and add them up, keeping in mind the CW and CCW sign convention.
- Calculate the moment caused by the resultant of the system of forces about that point.

The Principle of Moments tells us that they will be the same. Now there's a very important "alternate view" of the Principle of Moments. Let's say I have a force like the 60 N force you saw in the final example of the Moments module:

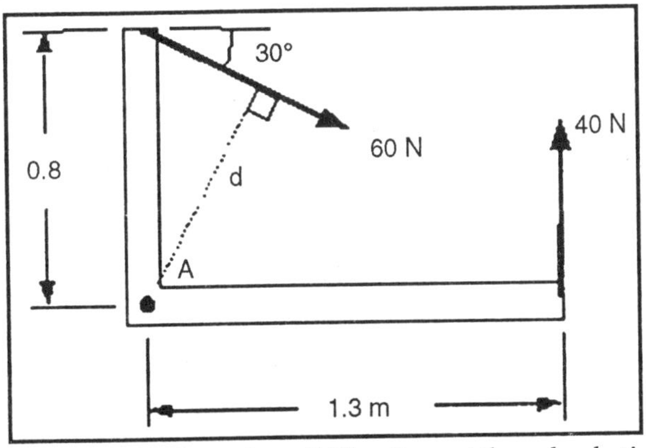

You'll recall that the difficulty with calculating the moment caused by the 60 N force about point A was that you had to calculate the moment arm using trigonometry. But if the 60 N force is resolved into its x and y components, then isn't the original 60 N force nothing more than a resultant of those two rectangular components?

If true (which it is...) then the Principle of Moments tells us that themoment of the 60 N force about point A will be equal to the moment we would get by calculating the moments due to it's components and then simply adding them up!

From the example, the x and y components of the 60 N force are calculated:

$Fx = 60 \cos 30 = 51.96$ N

$Fy = -60 \sin 30 = -30$ N (that's 30 N pointing downward, assuming a typical x-y axis)

So the equivalent system is as shown:

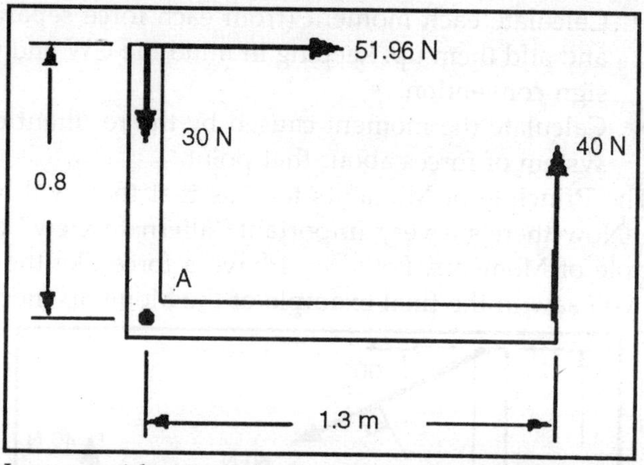

Now, considering only the component forces, look at the moment each one causes about point A (note the sign convention is ONLY due to whether the force causes a CW or CCW moment!)

$M_{A/51.96} = -51.96 \times 0.8 = -41.6$ N-m

$M_{A/30} = 30 \times 0 = 0$ N-m (the line of action passes through point A, so there is NO moment!)

Note that the moment arms are found very easily from the dimensions given in the problem. Now, when these moments are added together, the total moment turns out to be:

$M_A = -41.6 + 0 = -41.6$ N-m

Which is exactly the same as the moment calculated in the example from the previous module, where the moment arm for the resultant had to be calculated!

IMPORTANT!

You're going to see that for the vast majority of problems where you have to calculate a moment, the process will be greatly simplified by resolving a force into components and applying the Principle of Moments. In quite a few problems, you simply will not have enough information to calculate the moment any other way.

For this, let's determine the moment that the 150 N force creates about point A. Now before we do, just mentally think about the amount of effort you would have to put into finding the moment arm if you had to trigonometrically determine the distance between point A and the line of action of the force (remember, that's the distance that is perpendicular from that line of action!). Fortunately for us, we have Varignon's Theorem to work with. First, we resolve the 150 N force into x- and y-components (it will be helpful for you to sketch this out on paper, so you can visualize it better):

Fx = 150 cos(30) = 129.9 N

Fy = 150 sin(30) = 75 N

Using Varignon's Theorem, we have:

M_A = (-129.9 N × 120 mm) + (75 N × 350 mm) = 10,662 N-mm or 10.662 N-m

RESULTANT OF COPLANAR FORCES

There are many ways in which forces can be manipulated. It is often easier to work with a large, complicated system of forces by reducing it to an ever decreasing number of smaller problems. This is called the "resolution" of forces or force systems. This is one way to simplify what may otherwise seem to be an impossible system of forces acting on a body. Certain systems of forces are easier to resolve than others. Coplanar force systems have all the forces acting in in one plane. They may be concurrent, parallel, non-concurrent or non-parallel. All of these systems can be resolved by using graphic statics or algebra.

However, all of the individual vectors might not acutally be in contact with the common point. These are the most simple force systems to resolve with any one of many graphical or algebraic options.

A parallel coplanar force system consists of two or more forces whose lines of action are ALL parallel. This is commonly the situation when simple beams are analysed under gravity loads. These can be solved graphically, but are combined most easily using algebraic methods.

The last example is of a non-concurrent and non-parallel system. This consists of a number of vectors that do not meet at a single point and none of them are parallel. These systems are essentially a jumble of forces and take considerable care to resolve.

Almost any system of known forces can be resolved into a single force called a resultant force or simply a Resultant. Theresultant is a representative force which has the same effect on the body as the group of forces it replaces. (A couple is an exception to this) It, as one single force, can represent any number of forces and is very useful when resolving multiple groups of forces.

One can progressively resolve pairs or small groups of forces into resultants. Then another resultant of the resultants can be found and so on until all of the forces have been combined into one force. This is one way to save time with the tedious "bookkeeping" involved with a large number of individual forces. Resultants can be determined both graphically and algebraically.

It is often convenient to decompose a single force into two distinct forces. These forces, when acting together, have the same external effect on a body as the original force. They are known as components. Finding the components of a force can be viewed as the converse of finding a resultant. There are an infinate number of components to any single force. And, the correct choice of the pair to represent a force depends upon the most convenient geometry. For simplicity, the most convenient is often the coordinate axis of a structure.

These are known as the rectangular components of a force. Rectangular components can be thought of as the two sides of a right angle which are at ninety degrees to each other. The resultant of these components is the hypotenuse of the triangle. The rectangular components for any force can be found with trigonometrical relationships: $F_x = F \cos \theta$, $F_y = F \sin \theta$ (where theta is the angle). There are a few geometric relationships that seem common in general building practice in North America. These relationships relate to roof pitches, stair pitches, and common slopes or relationships between truss members.

Any concurrent set of forces, not in equilibrium, can be put into a state of equilibrium by a single force. This force is called the Equilibrant. It is equal in magnitude, opposite in sense and co-linear with the resultant. When this force is added to the force system, the sum of all of the forces is equal to zero.

Graphic Statics and graphical methods of force resolution were developed before the turn of the century by Karl Culmann. They were the only methods of structural analysis for many years. These methods can help to develop an intuitive understanding of the action of the forces. Today, the Algebraic Method is considered to be more applicable to structural design. Despite this, graphical methods are a very easy way to get a quick answer for a structural design problem and can aid in the determination of structural form.

Chapter 3

Equilibrium

CONCEPT OF EQUILIBRIUM

Equilibrium is the state of an object when no part of the object is accelerating. Two conditions for equilibrium are that the net force acting on the object is zero, and the net torque acting on the object is zero.

Architectural structures are normally stationary. Most clients, building officials and designers prefer that their structures remain static rather than move dynamically. There are specific loading conditions which are dynamic loads, but in each and every case a return to a stable and static state is desireable. Such a condition is known as equilibrium.

Various states of static equilibrium are experinced throughout one's life. Think of the "teeter-totter" at a playground or of a game of "tug-of-war." In the first case, two or more individuals sit upon a board which has been fixed to a fulcrum which allows rotation. If each of the individuals on the teeter totter weight exaclty the same amount and sit at exactly the same distance from the fulcrum the teeter-totter will not move. A state of equilibrium has been achieved. The two will remain at rest until an action takes them out of equilibrium.

Such an action could be the addition of another person to the system or it could be that one of the original two would change their position slightly. In either case, the teeter-totter would most likely swing to one side and rest upon the ground. A new state of equilibrium would have been found.

In this case, another boy climbed on behind the one

already sitting on the right. In order to put the system back into equilibrium the girl on the left had to move backwards along the board until she was far away. The moment two boys created around the fulcrum doubled when the second boy climbed on. The girl knew that the only way that she could increase the magnitude of the moment she created would be to increase the moment arm.

A structure is in equilibrium when all forces or moments acting upon it are balanced. This means that each and every force acting upon a body, or part of the body, is resisted by either another equal and opposite force or set of forces whose net result is zero.

Issac Newton addressed this issue when he noted that a body is at rest will remain at rest until acted upon by an external force. Every structure that can be seen to remain standing on a daily basis is in equilibrium; it is at rest and each of its members, combination of its members or any part of a member that is supporting a load are also at rest. There is a net result of zero in all directions for all of the applied loads and reactions.

EXPLANATION

The intersection of the supply and demand curve establishes equilibrium. Equilibrium prices are prices that equate buyers' willingness to pay for the last units of output (prices along demand curve) with sellers' willingness to sell (cost represented by the supply curve) the last units of output. Thus, prices are said to represent marginal evaluations of goods and services. At this equilibrium price, there is neither a shortage or surplus in the market.

To better understand the concept of market equilibrium, suppose sellers produced Q_1. To produce this level of output, sellers would have to receive a price equal to P_s. Sellers must receive P_s to cover the cost to produce the last units of output (remember the supply curve is an aggregate marginal cost curve). Buyers, however, are willing to pay only P_d for Q_1. The sellers will accumulate a surplus of output unless prices (and output) are reduced.

What if sellers produced Q_2? To produce this level of output, sellers would only need to charge a price equal to P_s. Buyers, however, are willing to pay up to P_d for Q_2. When consumers are willing to pay higher prices for additional output, but cannot buy the output, a shortage exists.

EXERCISES

- An object acted on by three forces moves with constant velocity. One force acting on the object is in the positive x direction and has a magnitude of 6.5 N; a second force has a magnitude of 4.4 N and points in the negative y direction. Find the magnitude of the third force acting on the object.
- A 10-kg sign is suspended by three ropes, each supporting an equal portion of the sign's weight. The two end ropes make an angle of 70º to the horizontal. What is the tension on each of the ropes?
- A 15 m rope is strung horizontally two poles. When a 47 kg mass is hung at the mid-point the rope sags 11°. What is the tension in the rope?
- A 29 kg sign is supported by a vertical chain at each end. A strong wind blows and the chains make a 30 degree angle with the vertical. What force does each chain exert?
- A 32.53 N light hangs at equilibrium from two cables at angles 78.3 and 60.2 with respect to the ceiling. What is the tension in the first cable?
- Two blocks on a ramp are connected by a string and tied to a wall. The mass of the lower block is 1.0 kg; the mass of the upper block is 2.0 kg. Given that the angle of the incline is 31 degrees, find the tensions in,
  - The string connecting the two blocks
  - The string tied to the wall.
- A lamp weighs 150 N and is supported equally by two wires that form an angle of 120 degrees with each other.
  - What is the tension in each of these wires?

- If the angle between the wires is reduced to 90 degrees, what is the new tension in each wire?

As the angle between the wires decrease, what happens to the force in the wires?

FREE BODY DIAGRAM

One of the most useful aids for solving a statics problem is the free body diagram (FBD). A free body diagram is a graphic, symbolic representation of the body (structure, element or piece of an element) with all connecting "parts" removed. The "parts" which have been removed are all of the real physical aspects of the structure.

The body is represented by a simple line. Every "part" which has been removed is replaced by the external forces representing the internal forces present where that "part" connects with the other member in the FBD. Loads, distributed or concentrated, are removed and replaced with representational force systems.

Everything that is needed to solve a force system is included on the FBD. Free body diagrams may not seem necessary in the relatively simple current applications, but as problems become more complex, their usefulness increases.

The following is the process for determining the reaction at the wall for a cantilever beam. A FBD is first drawn of the beam. Next, cut the beam free from the wall and replace the wall with the forces that were supporting the beam at the wall before it was cut free. These forces are unknown, but they are the only forces that can keep the beam in equilibrium. They are identical to the internal forces in the beam at that point before it was cut.

A fixed support will resist translation in all directions and rotation (moment). The FBD must show all of these directions. The principles of equilibrium can always be used to solve a FBD. In the FBD above Sum Fy = 2K and Sum Fx = 0. The 2K forces (load and vertical reaction force) cause a counter-clockwise couple of 10 K-FT which must be resisted by a moment on the end of the cut section of 10 K-FT acting in a clockwise direction.

The following are three different systems which all have one 100 pound load and one 150 pound load actin on them at exactly the same point.Note how the internal arrangements of the three has been removed. The internal arrangement does not matter for thé determination of the supporting reactions.

DETERMINATION OF UNKNOWN FORCES

A force can be resolved into an infinite number of combinations of components by the parallelogram method. The most useful components are the two components that are parallel to the X and Y axes. These are known as the rectangular components.

The solution of many problems is greatly simplified by using the rectangular components. These are usually noted as Fx, for forces parallel to the X axis, and Fy, for forces parallel to the Y axis. They are also known simply as H, for horizontal, and V, for vertical.

The rectangular components may be determined graphically, where the force is shown as a vector, or algebraically. In order to resolve a vector into its components, Fx = F cos or Fy = F sin, one must know at least two items of the six geometric descriptors of a triangle (the lengths of the sides and the three angles).

Conversely, if the magnitude of the rectangular components are known, the resultant can be found with the Pythagorean Theorem, F = SQRT (Fx^2 + Fy^2). The direction of the resultant may be determined by trigonometry knowing that the tan is = opposite side/adjacent side.

The solutions to many problems are greatly simplified by resolving multiple forces into a resultant or by finding the rectangular components of a force system.

In summary:

Fx = F cos

Fy = F sin

F = SQRT (Fx^2 + Fy^2)

tan = Fx /Fy = opposite side/adjacent side

The resultant of any number of concurrent forces can be found by resolving each force into its rectangular components and then adding the components algebraically. Remember, the sum of Fx = 0 and Fy = 0. The resulting numbers will be the compontents of the resultant. From these, the resultant can be found as before by using the Pythagorean Theorem in which R = SQRT (Fx^2 + Fy^2) and the formula for tan = Fy /Fx.

ALGEBRAIC FORCE RESOLUTION

Given: Four concurrent forces with the magnitudes and geometry shown.

Determine:

a. The V and H components of each of the forces
b. The resultant algebraically
c. The equilibrant.

Solution:

a. First break each of the forces into its horizontal and vertical components, either by inspection or using the algebraic method. The components can neatly be entered one by one into a chart similar to the one shown below.

Force Fx Fy:
60 - -
50 - -
100 - -
20 - -

The 20k force is the most straightforward, so enter its components into the chart first. Be careful to enter the correct sign for each component or the overall result will be incorrect.

Force Fx Fy:
60 - -
50 - -
100 - -
20 0 -20

The 60k force can also be resolved by observation because of its 3-4-5 geometry. Fx = 4/5(60k) = 48k and Fy = 3/5(60) = 36. Enter these numbers into the chart.

Force Fx Fy:
60 48 36
50 - -

100 - -

20 0 -20

The last two forces are most efficiently resolved using the algebraic equations:

Fx = F cos ψ

Fy = F sin ψ

So, for the 50k force, its components are:

Fx = -50k cos 30 = -50(.866) = -43.3k

Fy = 50k sin 30 = 50(.5) = 25k

and for the 100k force

Fx = -100k cos 45 = -100(.707) = -70.7k

Fy = -100k sin 45 = -100(.707) = -70.7k

Use these components to complete the chart.

Force Fx Fy:

60 48 36

50 -43.3 25

100 -70.7 -70.7

20 0 -20

b. The components of the resultant are equal to the sum of the columns of the completed chart

Force Fx Fy:

60 48 36

50 -43.3 25

100 -70.7 -70.7

20 0 -20

R -66.7 -29.7

The resultant is found using the pythagorean theorem:

= SQRT (sum Fx^2 + sum Fy^2)

= SQRT (-66^2 + -29.7^2)

= SQRT (4356 + 882)

= SQRT (5238)

= 72.4k

Its angle can be found by applying

tan ψ = opposite side [Fy]/adjacent side [Fx]

tan ψ = -29.7/-66

tan ψ =.45

ψ = arctan.45

ψ = 24.2 degrees below the x axis down and to the left

The quadrant can be found by observation using the signs of the components of the resultant. In this case, both Fx and Fy are negative, so the resultant will lie in the lower left quadrant.

The equilibrant will be the inverse of the resultant. Its magnitude will be the same, 72.4k, but it will act at an angle 24.2 degrees above the x axis, up and to the right.

PARALLEL FORCE SYSTEMS

Given: The resultant (this includes the Magnitude, Direction and Sense, and a Point on its Line of Action); AND, the force(s) required to put this system into equilibrium.

Solution: Since there are only vertical forces in this system, only the sum of the vertical forces will have to be determined in order to find the magnitude of the resultant of the system.

Sum Fy = Resultant = -800# + 600# - 1200# -400# = -1800#

The resultant is -1800# or 1800# down. The location of the resultant must be determined through the moment equilibrium equation. In this case take the sum of the moments around A for simplicity, although any point would yield the same result.

Sum MA = 1800# x = –600#(3ft) + 1200#(5ft) + 400# (9ft)

= –1800#ft + 6000#ft + 3600#ft

= 7800#ft

x = 7800#ft/1800# = 4.33ft

This distance can be checked by taking the moment around another point. (Though not the point through which the resultant acts) Take, for example, the other end of the beam. The distance for the resultant should be

9ft (total length of the beam) - 4.33ft (distance from A)

= 4.67ft (distance from the other end)

Sum M = -1800#(4.67ft) = -1200#(4ft) + 600#(6ft) - 800#(9ft)

–8406#ft = -4800#ft + 3600#ft - 7200#ft

–8406#ft = -8400#ft *

Since the distance 'd' is a repeating decimal, the amount of discrepence will depend upon the number of decimal places used. In this case, these numbers are sufficient to confirm the location of the resultant.

Knowing the magnitude, direction and location of the resultant, the equilibrant is found by simply inversing the direction of the resultant. The equilibrant acts with a force of 1800# up at a point 4.33 ft to the right of point Á.

Chapter 4

Centre of Gravity

GRAVITY AND GRAVITATIONAL FORCE

GRAVITY

The terms gravity and gravitation are often used to explain the same thing, but there is a definite difference between the two.

Gravitation is the attractive force existing between any two objects that have mass. The force of gravitation pulls objects together.

Gravity is the gravitational force that occurs between the earth and other bodies. Gravity is the force acting to pull objects toward the earth.

Since gravitational force is happening to all matter (objects) in the universe, from the largest galaxies down to the smallest atoms, it is often called *universal gravitation*. Sir Isaac Newton was the first to fully recognize that the force holding any object to the earth is the same as the force holding the moon, the planets, and other heavenly bodies in their orbits.

GRAVITATION

Is actually a *very weak force*. The pull is too weak to be felt between two people and it is not even strong enough to pull together two lumps of lead placed right next to each other. It is only when one of the masses is the size of a planet that we can feel the force of gravity.

Geophysical interpretations from gravity surveys are based on the mutual attraction experienced between two masses as first expressed by Isaac Newton in his classic work *Philosophiae naturalis principa mathematica* (The mathematical principles of natural philosophy). Newton's law of gravitation states that the mutual attractive force between two point masses, *m1* and *m2*, is proportional to one over the square of the distance between them. The constant of proportionality is usually specified as *G*, the gravitational constant. Thus, we usually see the law of gravitation written as shown to the right where *F* is the force of attraction, *G* is the gravitational constant, and*r* is the distance between the two masses, *m1* and *m2*.

The huge gravitational force of our nearest star, the Sun, holds together the nine planets of our Solar System. The planets move through space at speeds that just balance the Sun's gravitational pull, so they are locked into a permanent circle (orbit) around the Sun. Moons orbit planets, and satellites and spacecraft orbit the Earth, in the same way. Satellites are not defying gravity in circling endlessly around the Earth, it is just that they are moving so fast around the

Gravity indicates gravitational force that occurs between the earth and other bodies. Gravity is the force acting to pull objects toward the earth.

Gravity is the force that holds us on the ground and causes objects to fall back to the ground after being thrown in the air.

The force holding objects to the earth's surface depends not only on the Earth's gravitational field but also on other factors, such as the Earth's rotation.

The Earth's gravitational pull extends out into space in all directions. The further you move away from the centre of the Earth the weaker the force becomes.

The measure of the force of gravity on an object is the *weight* of that object. Weight is measured in newtons (N). The weight of an object changes depending on its location in the universe.

DETERMINATION OF CENTRE OF GRAVITY

In general, determining the centre of gravity is a complicated procedure because the mass (and weight) may not be uniformly distributed throughout the object. The general case requires the use of calculus which we will discuss later. If the mass is uniformly distributed, the problem is greatly simplified. If the object has a line (or plane) of symmetry, the CG lies on the line of symmetry. For a solid block of uniform material, the centre of gravity is simply at the average location of the physical dimensions. (For a rectangular block, 50 × 20 × 10, the centre of gravity is at the point (25, 10, 5).) For a triangle of height h, the CG is at $h/3$, and for a semicircle of radius r, the cg is at $(4r/(3\pi))$ where π is ratio of the circumference of the circle to the diametre. There are tables of the location of the centre of gravity for many simple shapes in math and science books. The tables were generated by using the equation from calculus shown in the diagram.

For an object of general shape, there is a simple mechanical way to determine the centre of gravity:

If we just balance the object using a string or an edge, the point at which the object is balanced is the centre of gravity. (Just like balancing a pencil on your finger!)

Another, more complicated way, is a two-step method illustrated above. In Step 1, you hang the object from any point and you drop a weighted string from the same point. Draw a line on the object along the string. For Step 2, repeat the procedure from another point on the object You now have two lines drawn on the object which intersect. The centre of gravity is the point where the lines intersect. This procedure works well for irregularly shaped objects that are hard to balance.

If the mass of the object is not uniformly distributed, we must use calculus to determine centre of gravity. The centre of gravity can be determined from:

$$CG \times W = +'' \; x \, dw$$

where x is the distance from a reference line, dw is an increment of weight, and W is the total weight of the object. To evaluate the right side, we have to determine how the weight varies geometrically. From the weight equation, we know that:

$$w = m \times g$$

where m is the mass of the object, and g is the gravitational constant. In turn, the mass m of any object is equal to the density, ρ (rho), of the object times the volume, V:

$$m = r \times V$$

We can combine the last two equations:

$w = g \times \rho \times V$

then

$dw = g \times \rho \times dV$

$dw = g \times \rho(x,y,z) \times dx\ dy\ dz$

If we have a functional form for the mass distribution, we can solve the equation for the centre of gravity:

CG × $W = g \times$ +++" $x \times \rho(x,y,z)\ dx\ dy\ dz$

where +++" indicates a triple integral over dx, dy, and dz. If we don't know the functional form of the mass distribution, we can numerically integrate the equation using a spreadsheet. Divide the distance into a number of small volume segments and determining the average value of the weight/volume (density times gravity) over that small segment. Taking the sum of the average value of the weight/volume times the distance times the volume segment divided by the weight will produce the centre of gravity.

CENTRE OF MASS

The centre of mass (/gravity) of a body or a system of particles is the resultant of the weights of the individual particles making up the body or system.

UNIFORM BODIES

A uniform body is a body whose density is the same throughout the body. Finding the centre of mass of uniform bodies is relatively straightforward.

A lamina is a 2-dimensional object. In other words, it is a flat object whose thickness is we can ignore.

If a body has a line of symmetry, the centre of mass will lie on this line. This fact is very useful. For example, the centre of mass of a circular lamina is at the centre of the circle, since the centre of mass is on each axis of symmetry and they all meet at the centre.

STANDARD RESULTS

- The centre of mass of a uniform semicircular lamina of radius r lies on the axis of symmetry at a distance of 4r/3p from the straight edge
- The centre of mass of a uniform solid right circular cone of height h lies on the axis of symmetry at a distance of h/4 from the base
- The centre of mass of a uniform solid hemisphere of radius r lies on the axis of symmetry at a distance of 3r/8 from the base

USING INTEGRATION (M3)

When the centre of mass of a body cannot be found using the axis/axes of symmetry, it can be found by integration. Remember, from the definition of the centre of mass, it is the resultant of the weights of the individual particles making up the body. So we can take moments about any point.

Moment of the total = the total of moments.

In other words, (distance of centre of mass from O) × (weight of body) = the sum of: (the mass of each particle) × (the distance of each particle from O).

The "sum of" part can be replaced by an integral. This might sound a bit confusing at first, but with practise it becomes second nature. Another useful equation is: mass = density × volume

EXAMPLE

Find the position of the centre of mass of a uniform semi-circular lamina, radius r. We know that the centre of mass lies somewhere on the axis of symmetry of the semi-circle, although where on the axis of symmetry we do not know. We therefore divide the semi-circle into many tiny strips, each with thickness dx.

We know that the centre of mass lies on the red line, since this is a line of symmetry of the object. If we label this line the x-axis and introduce a y-axis as follows, this will allow is to use calculus:

Each strip is a distance of x from the y-axis. The mass of each strip is its volume × density. The volume of each strip is just its area, since it is a lamina and so has no depth. The area of each strip is dx × 2y.

We are told that the body is uniform, which means that the density is constant. It doesn't matter what the density is, lets call it r. So the mass of each strip is 2y rdx.

Now the total area of the semicircle is ½ pr^2 . So the total mass is ½ pr^2r. (distance of centre of mass from O) × (weight of body) = the sum of: (the mass of each particle) × (the distance of each particle from O). Therefore: (centre of mass) × ½ pr^2r = S 2y x r dx. The sum is from x = 0 to x = r. (centre of mass) × ½ pr^2r =2y x r dx.

The integral is with respect to x, so we must replace the y by a function of x.

By thinking about similar triangles, we can deduce that y/r = x/r, so y = x

Therefore:

(Centre of mass) × ½ $pr^2r = 2x^2$ r dx

(Centre of mass) × ½ $pr^2r = (2r^3 r)/3$

So centre of mass is a distance of 4r/3p from O, on the axis of symmetry.

MASS MOMENT OF INERTIA

Mass moment of inertia for a particle: The mass moment of inertia is one measure of the distribution of the m ıss of an object relative to a given axis.

As can be seen from the above equation, the mass moment of inertia has the units of mass times length squared. The mass moment of inertial should not be confused with the area moment of inertia which has units of length to the power four. Mass moments of inertia naturally appear in the equations of motion, and provide information on how difficult (how much inertia there is) it is rotate the particle around given axis.

The Mass Moment of Inertia of a solid measures the solid's ability to resist changes in rotational speed about a specific axis. The larger the Mass Moment of Inertia the smaller the angular acceleration about that axis for a given torque.

The mass moment of inertia depends on a reference axis, and is usually specified with two subscripts. This helps to provide clarity during three-dimensional motion where rotation can occur about multiple axes.

Mass moment of inertia for a rigid body: When calculating the mass moment of inertia for a rigid body, one thinks of the body as a sum of particles, each having a mass of *dm*. Integration is used to sum the moment of inertia of each *dm* to get the mass moment of inertia of body.

PARALLEL AXIS THEOREM

The swingweight of a racquet is measured by the Babolat RDC using an axis of rotation 10 cm from the butt. To find the swingweight about the axis used on the stroke requires application of the Parallel Axis Theorem.

Swingweight, also known as moment of inertia and rotational inertia, is the resistance to change in the speed of the rotation about the axis of rotation. High swingweight means that the racquet is hard to get rotating, but once it gets going it will not be pushed around so much on impact with the ball and will tend to produce better pace and spin. Swingweight is the infinite sum of all infinitely small mass elements times the square of their distance from the axis of rotation. The swingweight of the racquet about its mass centre will be called Ic. The distance of the mass centre (balance point of the racquet) from the axis of rotation for the known swingweight is r, and the racquet mass is M. Note that the published mass centre radius is measured from the balance point to the butt end, not to the axis of rotation which is 10 cm from the butt end. The distance of the mass centre from the new axis of rotation is r + x.

The Parallel Axis Theorem (from first year physics) tells us that the swingweight is the sum of the swingweight about the mass centre (Ic) plus the product of the mass (M) and the square of the distance (r) from the axis of rotation to the mass centre.

$$I = Ic + Mr2$$

So once we know the RDC swingweight about the 10 cm

axis, we can find the swingweight about any other parallel axis by another application of the Parallel Axis Theorem. The first step is finding the swingweight about an axis through the mass centre (Ic).

I = Ic + Mr2

=> Ic = I – Mr2

The same holds for the unknown swingweight, which is about a parallel axis that is a distance r + x from the mass centre:

I' = Ic + M(r + x)2

Now substitute what we've already found is equal to Ic:

I' = (I – Mr2) + M(r + x)2

= (I – Mr2) + M(r2 + 2rx + x2)

Simplifying, we get a general formula for finding the unknown swingweight (I') about a different axis (r + x):

I' = I + M(2rx + x2)

For the First Benchmark Condition (groundstroke), the axis of rotation for the stroke is at 7 cm from the handle end, so x is 3. And in the Second Benchmark Condition (serve), the axis is even farther away, and x is 5. The variable r is the published balance point minus 10 cm, but in our formulas we use r, which is the published balance point minus 7 or 5 cm according to which benchmark condition we are using, so we need to substitute in the above formula so we can use r instead of r. We will use another variable (a) to denote the axis used. *So*:

x = 10 - a

r = r + (10 – a) → r = r – (10 – a)

When we substitute in the above formula, we get:

I' = I + M(2rx + x2) = I + M[2(r – (10 - a))(10 – a) + (10 – a)2]

Now simplifying the expression in the brackets:

[2(r – (10 – a))(10 – a) + (10 – a)2]

= -2ar + 20a – 2a2 + 20r – 200 + 20a + 100 – 20a + a2

= -2ar + 20a – a2 + 20r – 100

= 20r – 2ar – 100 + 20a – a2

= 2r(10 – a) – (10 - a)2

So in the formulas, the variable I, which represents the swingweight about the axis used on the stroke, where a is the distance from the butt to the axis, I10 is the RDC swingweight about the axis 10 cm from the butt, and r is the distance in cm from the mass centre (balance point) to the axis used in the stroke:

I = Ia = I10 + M*[2*(10 – a)*r – (10 – a)^2]

a = distance in cm from the butt to the axis of rotation used in the stroke.

I10 = RDC swingweight in kg.cm2 about an axis 10 cm from the butt.

r = distance in cm from mass centre (balance point) to axis used in the stroke.

Chapter 5

Kinematics

KINEMATICS OF PARTICLES

DEFINITION

Kinematics is the study of how things move. Here, we are interested in the motion of normal objects in our world. A normal object is visible, has edges, and has a location that can be expressed with (x, y, z) coordinates. We will not be discussing the motion of atomic particles or of light.

We will create a vocabulary and a group of mathematical methods that will describe this ordinary motion. It is important to understand that we will be only developing a language used for describing motion here. We will not be concerned with what is causing the objects to change their motion, or more correctly, their momentums. In other words, we are not concerned with the action of forces within this topic.

3D THEORY OF KINEMATICS

Representing Motion in 3D

To model motion of a body, we can use a transform (T) to transform each point relative to the body to a position in world coordinates. If we want to model both liniear and angular (rotational) motion then we need to use a 4x4 matrix to represent the transform.

If we only need to model rotation then, we can reduce this to either use a 3x3 matrix or a quaternion.

Modelling rotations in 3D is not always intuative. For

instance, a rotation of 90 degrees about the x-axis followed by a rotation of 90 degrees about the y-axis is equivilant to a rotation of 90 degrees about the z-axis.

Frameworks, Chains

A lot of mechanical objects in the real world consist of solid sections connected by joints.

Examples of this might be:

- Creatures such as humans and animals.
- Car Suspension
- Robot arms
- Ropes, string and Chains.

Where you have something like a robot arm, each section of the arm can pivot, bend and/or rotate from the previous stage. These sections and joints are known as a'chain', if we are talking about creatures then it could be referred to as a skeleton. Similarly in biological terms a movable section can be referred to as a bone, and the attachments between each bones as joints.

Motions of chains can be specified in terms of translations and rotations.

"*Forward Kinematics*: From the amounts of rotation and bending of each joint in an arm, for example, the position of the hand can be calculated.

"*Inverse Kinematics*: If the hand is moved, the rotation and bending of the arm is calculated, in accordance with the length and joint properties of each section of the arm.

Forward Kinematics

This Chain is represented here, without any information about its actual shape, what we are interested in is the relative positions of all the joints.

The first link is the base (or"anchor point" in nautical terms) is the end of the chain which always has a fixed or known position.

From there, each link can be rotated (possibly in different dimensions depending on the degree of freedom of the joint).

The other important factor is the distance between the

joints, this can be represented as a translation in a TransformGroup. This transform group can also hold the rotation of that link. So Forward Kinematics allows one to find the position of the endpoint (and all the points in-between) relative to the base. This is easy to do, just multiply by the TransformGroup matrixes in turn.

Inverse Kinematics

For inverse kinematics (IK), the position of the end point is known, and we need to find the angles of the joints. This is a much harder problem, there may be many possible answers, or there may not be a set of angles that would reach to that point. Although this is hard to do, it is useful, for example, when posing a model it is useful to be able to drag the extremities with a mouse and have the arms/links position themselves appropriately.

There are two approaches to solving inverse kinematics:

- *Analytical*: Requires a lot of trigonometry or matrix algebra
- *Iterative*: Better if there are lots of links and degrees of freedom.

IK - Analytical Approach

If there are only two or three links then it may be possible to solve it analytically. One possibly might be to draw out the arm with the angles shown on it, then solve for the angles using geometry, a good knowledge of trig identities would help here. The problem is that this is not really very general approach.

Another analytical approach is to represent each links rotation and translation by a matrix. The end point is then given by all these matrixes multiplied together, so we just need to solve this matrix equation. Then find what rotation each matrix represents. There may be many solutions or there may not be any solutions. In other words there are lots of ways to reach to a given point, or it may be out of reach.

If there are many solutions, then you might need to apply additional constraints. For instance, human joints can only bend within certain limits.

IK - Iterative -approach

This is a more general approach for programming complex chains. Start off with the joints in any position, then move each of the joints in turn, so that each movement takes the endpoint toward the target.

Starting with the joint nearest the end point, rotate the joint so that the current end point moves toward the required end point. Then do the same with the next joint toward the base and so on until the base is rotated. Then keep repeating this, until the end point is close enough to the required end point or if further iterations are not moving it closer to the required point.

It may be possible to have a more realistic strategy than this, for instance, if I am using my arm to pick up an object then, if the object is a long way away, I will move the bigger joints in the arm, then as the hand gets closer the smaller joints of the hand are used for the fine adjustments.

The angle of rotation for each joint is found by taking the dot product of the vectors from the joint to the current point and from the joint to the desired end point. Then taking the arcsin of this dot product.

Human Animation

This can be represented by a scenegraph tree of transform groups

HumanoidRoot: sacrum
sacroiliac: pelvis
| l_hip: l_thigh
| l_knee: l_calf
| l_ankle: l_hindfoot
| l_subtalar: l_midproximal
| l_midtarsal: l_middistal
| l_metatarsal: l_forefoot
| r_hip: r_thigh
| r_knee: r_calf
| r_ankle: r_hindfoot
| r_subtalar: r_midproximal
| r_midtarsal: r_middistal

| r_metatarsal: r_forefoot
vl5: l5
vl4: l4
vl3: l3
vl2: l2
vl1: l1
vt12: t12
<snip>...

This skeleton is then combined with a physical model of the human/creature to produce the animation

KINEMATIC EQUATIONS

Ima Hurryin is approaching a stoplight moving with a velocity of +30.0 m/s. The light turns yellow, and Ima applies the brakes and skids to a stop. If Ima's acceleration is -8.00 m/s2, then determine the displacement of the car during the skidding process. (Note that the direction of the velocity and the acceleration vectors are denoted by a + and a - sign.)

The solution to this problem begins by the construction of an informative diagram of the physical situation. This is shown below. The second step involves the identification and listing of known information in variable form. Note that the vf value can be inferred to be 0 m/s since Ima's car comes to a stop. The initial velocity (vi) of the car is +30.0 m/s since this is the velocity at the beginning of the motion (the skidding motion). And the acceleration (a) of the car is given as - 8.00 m/s2. (Always pay careful attention to the + and - signs for the given quantities.) The next step of the strategy involves the listing of the unknown (or desired) information in variable form. In this case, the problem requests information about the displacement of the car. So d is the unknown quantity. The results of the first three steps are shown in the table below.

vi = +30.0 m/s
vf = 0 m/s
a = - 8.00 m/s2d = ??

The next step of the strategy involves identifying a kinematic equation which would allow you to determine the unknown quantity. There are four kinematic equations to choose from. In general, you will always choose the equation which contains the three known and the one unknown variable.

In this specific case, the three known variables and the one unknown variable are vf, vi, a, and d. Thus, you will look for an equation which has these four variables listed in it.

Once the equation is identified and written down, the next step of the strategy involves substituting known values into the equation and using proper algebraic steps to solve for the unknown information. This step is shown below.

(0 m/s)2 = (30.0 m/s)2 + 2*(-8.00 m/s2)*d
0 m2/s2 = 900 m2/s2 + (-16.0 m/s2)*d
(16.0 m/s2)*d = 900 m2/s2 - 0 m2/s2
(16.0 m/s2)*d = 900 m2/s2
d = (900 m2/s2)/(16.0 m/s2)
d = (900 m2/s2)/(16.0 m/s2)
d = 56.3 m

The solution above reveals that the car will skid a distance of 56.3 metres. (Note that this value is rounded to the third digit.)

The last step of the problem-solving strategy involves checking the answer to assure that it is both reasonable and accurate. The value seems reasonable enough. It takes a car a considerable distance to skid from 30.0 m/s (approximately 65 mi/hr) to a stop. The calculated distance is approximately one-half a football field, making this a very reasonable skidding distance. Checking for accuracy involves substituting the calculated value back into the equation for displacement and insuring that the left side of the equation is equal to the right side of the equation. Indeed it is!

Example B

Ben Rushin is is waiting at a stoplight. When it finally turns green, Ben accelerated from rest at a rate of a 6.00 m/s2 for a time of 4.10 seconds. Determine the displacement of Ben's car during this time period.

The second step of the strategyinvolves the identification and listing of known information in variable form. Note that thevi value can be inferred to be 0 m/s since Ben's car is initially at rest. The acceleration (a) of the car is 6.00 m/s2. And the time (t) is given as 4.10 s. The next step of the strategyinvolves the listing of the unknown (or desired) information in variable form. In this case, the problem requests information about the displacement of the car. So d is the unknown information.

The next step of the strategy involves identifying a kinematic equation which would allow you to determine the unknown quantity. There are four kinematic equations to choose from. Again, you will always search for an equation which contains the three known variables and the one unknown variable. In this specific case, the three known variables and the one unknown variable are t, vi, a, and d. An inspection of the four equations above reveals that the equation on the top left contains all four variables.

Once it is identified and written down, the next step of the strategy involves substituting known values into the equation and using proper algebraic steps to solve for the unknown information. This step is shown below.

d = (0 m/s)*(4.1 s) + 0.5*(6.00 m/s2)*(4.10 s)2

d = (0 m) + 0.5*(6.00 m/s2)*(16.81 s2)

d = 0 m + 50.43 m

d = 50.4 m

The solution above reveals that the car will travel a distance of 50.4 metres. (Note that this value is rounded to the third digit.)

The last step of the problem-solving strategy involves checking the answer to assure that it is both reasonable and accurate. The value seems reasonable enough. A car with an acceleration of 6.00 m/s/s will reach a speed of approximately 24 m/s (approximately 50 mi/hr) in 4.10 s. The distance over which such a car would be displaced during this time period would be approximately one-half a football field, making this a very reasonable distance. Checking for accuracy involves substituting the calculated value back into the equation for displacement and insuring that the left side of the equation is equal to the right side of the equation. Indeed it is!

The two example problems above illustrate how the kinematic equations can be combined with a simple problem-solving strategy to predict unknown motion parametres for a moving object. Provided that three motion parametres are known, any of the remaining values can be determined. In the next part of Lesson 6, we will see how this strategy can be applied to free fall situations. Or if interested, you can try some practice problems and check your answer against the given solutions.

KINEMATIC EQUATIONS AND FREE FALL

As mentioned a free-falling object is an object which is falling under the sole influence of gravity. That is to say that any object which is moving and being acted upon only be the force of gravity is said to be"in a state of free fall." Such an object will experience a downward acceleration of 9.8 m/s/s. Whether the object is falling downward or rising upward towards its peak, if it is under the sole influence of gravity, then its acceleration value is 9.8 m/s/s.

The symbols in the above equation have a specific meaning: the symbol d stands for the displacement; the symbol t stands for the time; the symbol a stands for the acceleration of the object; the symbol vi stands for the initial velocity value; and the symbol vf stands for the final velocity.

There are a few conceptual characteristics of free fall motion which will be of value when using the equations to analyse free fall motion.

These concepts are described as follows:

- An object in free fall experiences an acceleration of -9.8 m/s/s. (The - sign indicates a downward acceleration.) Whether explicitly stated or not, the value of the acceleration in the kinematic equations is -9.8 m/s/s for any freely falling object.
- If an object is merely dropped (as opposed to being thrown) from an elevated height, then the initial velocity of the object is 0 m/s.
- If an object is projected upwards in a perfectly vertical direction, then it will slow down as it rises upward. The instant at which it reaches the peak of its trajectory, its velocity is 0 m/s. This value can be

used as one of the motion parametres in the kinematic equations; for example, the final velocity (vf) after traveling to the peak would be assigned a value of 0 m/s.

- If an object is projected upwards in a perfectly vertical direction, then the velocity at which it is projected is equal in magnitude and opposite in sign to the velocity which it has when it returns to the same height. That is, a ball projected vertically with an upward velocity of +30 m/s will have a downward velocity of -30 m/s when it returns to the same height.

These four principles and the four kinematic equations can be combined to solve problems involving the motion of free falling objects. The two examples below illustrate application of free fall principles to kinematic problem-solving. In each example, theproblem-solving strategy which was introduced earlier in this lesson will be utilized.

Example

Luke Autbeloe drops a pile of roof shingles from the top of a roof located 8.52 metres above the ground. Determine the time required for the shingles to reach the ground.

The solution to this problem begins by the construction of an informative diagram of the physical situation. This is shown below. The second step involves the identification and listing of known information in variable form. You might note that in the statement of the problem, there is only one piece of numerical information explicitly stated: 8.52 metres. The displacement (d) of the shingles is -8.52 m. (The - sign indicates that the displacement is downward). The remaining information must be extracted from the problem statement based upon your understanding of the above principles. For example, the vi value can be inferred to be 0 m/s since the shingles are dropped (released from rest). And the acceleration (a) of the shingles can be inferred to be -9.8 m/s2 since the shingles are free-falling. (Always pay careful attention to the + and - signs for the given quantities.) The next step of the solution involves the listing of the unknown (or desired) information in variable form. In this case, the problem requests information about the time of fall.

There are four kinematic equations to choose from. In general, you will always choose the equation which contains the three known and the one unknown variable. In this specific case, the three known variables and the one unknown variable are d, vi, a, and t. Thus, you will look for an equation which has these four variables listed in it.

Once the equation is identified and written down, the next step involves substituting known values into the equation and using proper algebraic steps to solve for the unknown information. This step is shown below.

-8.52 m = (0 m/s)*(t) + 0.5*(-9.8 m/s2)*(t)2

-8.52 m = (0 m) *(t) + (-4.9 m/s2)*(t)2

-8.52 m = (-4.9 m/s2)*(t)2

(-8.52 m)/(-4.9 m/s2) = t2

1.739 s2 = t2

t = 1.32 s

The solution above reveals that the shingles will fall for a time of 1.32 seconds before hitting the ground. (Note that this value is rounded to the third digit.)

The last step of the problem-solving strategy involves checking the answer to assure that it is both reasonable and accurate. The value seems reasonable enough. The shingles are falling a distance of approximately 10 yards (1 metre is pretty close to 1 yard); it seems that an answer between 1 and 2 seconds would be highly reasonable. The calculated time easily falls within this range of reasonability. Checking for accuracy involves substituting the calculated value back into the equation for time and insuring that the left side of the equation is equal to the right side of the equation.

Example

Rex Things throws his mother's crystal vase vertically upwards with an initial velocity of 26.2 m/s. Determine the height to which the vase will rise above its initial height.

Once more, the solution to this problem begins by the construction of an informative diagram of the physical situation. This is shown below. The second step involves the identification and listing of known information in variable form. You might note that in the statement of the problem, there is only one piece of numerical information explicitly stated: 26.2 m/s. The initial velocity (vi) of the vase is +26.2 m/s. (The + sign indicates that the initial velocity is an upwards velocity).

The remaining information must be extracted from the problem statement based upon your understanding of the above principles. Note that the vf value can be inferred to be 0 m/s since the final state of the vase is the peak of its trajectory. The acceleration (a) of the vase is -9.8 m/s2. The next step involves the listing of the unknown (or desired) information in variable form.

Once the equation is identified and written down, the next step involves substituting known values into the equation and using proper algebraic steps to solve for the unknown information. This step is shown below.

(0 m/s)2 = (26.2 m/s)2 + 2*(-9.8m/s2)*d
0 m2/s2 = 686.44 m2/s2 + (-19.6 m/s2)*d
(-19.6 m/s2)*d = 0 m2/s2 -686.44 m2/s2
(-19.6 m/s2)*d = -686.44 m2/s2
d = (-686.44 m2/s2)/(-19.6 m/s2)
d = 35.0 m

The solution above reveals that the vase will travel upwards for a displacement of 35.0 metres before reaching its peak. (Note that this value is rounded to the third digit.)

The last step of the problem-solving strategy involves checking the answer to assure that it is both reasonable and accurate. The value seems reasonable enough. The vase is thrown with a speed of approximately 50 mi/hr (merely approximate 1 m/s to be equivalent to 2 mi/hr). Such a throw will never make it further than one football field in height (approximately 100 m), yet will surely make it past the 10-yard line (approximately 10 metres). The calculated answer certainly falls within this range of reasonability. Checking for accuracy involves substituting the calculated value back into the equation for displacement and insuring that the left side of the equation is equal to the right side of the equation. Indeed it is!

Kinematic equations provide a useful means of determining the value of an unknown motion parametre if three motion parametres are known. In the case of a free-fall motion, the acceleration is often known.

And in many cases, another motion parametre can be inferred through a solid knowledge of some basic kinematic principles.

INTEGRATED MOTION SOLUTION FOR ROBOT CONTROL

Using robotics can greatly improve your machine's efficiency and performance. However, integrating robotics into your machine can be complicated and time consuming, often requiring a mix of disparate hardware and software components from different vendors.

Rockwell Automation has made robot control much easier - with Kinematics seamlessly integrated into the Kinetix Integrated Motion Solution.

Kinematics provides a single, integrated solution for multiple disciplines, eliminating additional robot and safety controllers, software, and special custom function blocks typically needed when incorporating a robot into an application. This significantly reduces the time and costs for both design and programming.

OEMs and end users can easily programme the robot in simple Cartesian coordinates, while the controller handles Kinematics transformations. With RSLogix 5000 software, you control 2 and 3 axis articulated independent/dependent, SCARA, H-bot and delta geometry robots natively in standard and safety ControlLogix controllers.

Because of the way the Kinematics transformation is implemented in RSLogix 5000 and the ControlLogix controllers, you can enable the Kinematics transformation with a newly added instruction and then just re-use the extensive and familiar set of motion instructions to programme the robot and the motion components of the application.

PROJECTILES

The first internal combustion engine was the gun, its piston the projectile that left its muzzle, propelled by the gases at high pressure produced by the explosion of its charge. The intention of the gunner was to send the projectile to a predetermined point, to execute its destructive or evil purpose.

The importance of this process in the affairs of mankind has led to the study of the paths of projectiles, or *ballistics*. At school, ballistics is a problem in uniformly accelerated motion, and the trajectory is a parabola, determined by the initial velocity and the angle of projection. In practice, this description falls far short of reality because of the influence of the atmosphere on the motion of the projectile. In studying the consequences of air resistance, we will come across many peculiar facts, and expert guidance in the matter of streamlining.

Suppose that we measure the velocity of a projectile in level flight at two points separated by a distance x, finding values V and v.

The average retarding force at velocity $(V + v)/2$ is then

$$F = m(V^2 - v^2)/2x$$

where m is the mass of the projectile.

This force is called *drag*. If the drag as a function of the velocity is known, then the trajectory of the projectile can be calculated, at least in principle. We require, therefore a means of measuring the velocity of a projectile.

A means was devised by Benjamin Robins (1707-1751), the *ballistic pendulum*, around 1842. The projectile is fired into a heavy pendulum, to which it transfers its momentum. From the swing and weight of the pendulum, the velocity of the projectile can be found. This is still an experiment performed in the Physics laboratory. Robins measured the velocity of the projectile at various distances from the gun, using the same charge each time.

He showed that the drag was roughly proportional to the square of the velocity, the first time this was done, but accurate results were impossible due to the lack of reproducibility of the muzzle velocity even when the same charge was used.

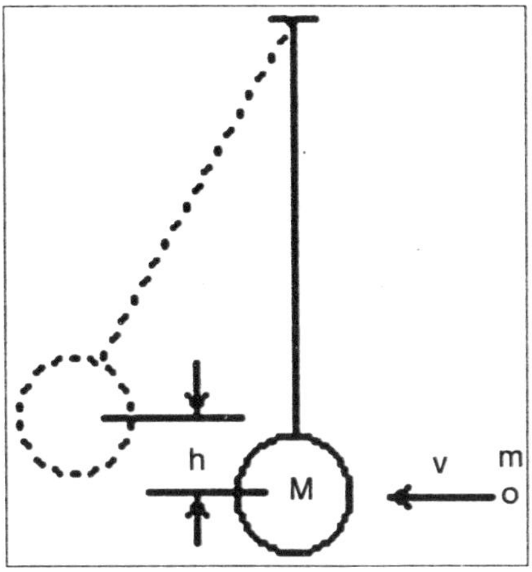

Fig: The Ballistic Pendulum

Hutton, at Woolwich Arsenal around 1812, suspended the gun as a pendulum itself. From its recoil, the muzzle velocity could be determined, so that accurate results for the drag as a function of velocity could be obtained.

Bashforth, also at Woolwich, devised an electrical timer that gave the time at which the projectile passed through each in a sequence of screens. From these times, the speed as a function of distance can be determined.

Around the time of the First World War, wind tunnel measurements began to measure the drag directly. So important were these measurements regarded that they were initially military secrets.

The drag on a projectile can be expressed as,

$$F = \rho d^2 v^2 f(v/c, vd/\nu).$$

Here, ρ is the density of the air, d is the diametre of the projectile, v its velocity, and the dimensionless quantity f(M,R) is called the *drag coefficient*, depending on the dimensionless parametres M = v/c, the Mach number, and vd/í, the Reynolds number, where í is the kinematic viscosity of the air, about 0.144 cm^2/s under normal conditions. When comparing drag coefficients, carefully note the defining equations, which may differ in the constants used.

For values of R below some critical value, on the order of

100 or 1000, the drag coefficient f = C/R = Cí/vd, so F = Cpídv, where C is some constant.

The drag is proportional to the velocity, and to the linear dimension of the projectile. This *viscous* regime is rather unimportant for projectiles, since even for a projectile as small as 5 mm diametre, the critical speed is below 2.8 m/s.

Above the critical value of R, f is roughly constant, say f = C′, and $F = C\rho d^2v^2$, proportional to the square of the velocity and the cross-sectional area of the projectile. The nature of the flow is shown in the Figure. It is the same whether the projectile is considered to move in still air, or whether the air is moving past the fixed projectile at the same speed. Streamline flow occurs where the streamlines are converging.

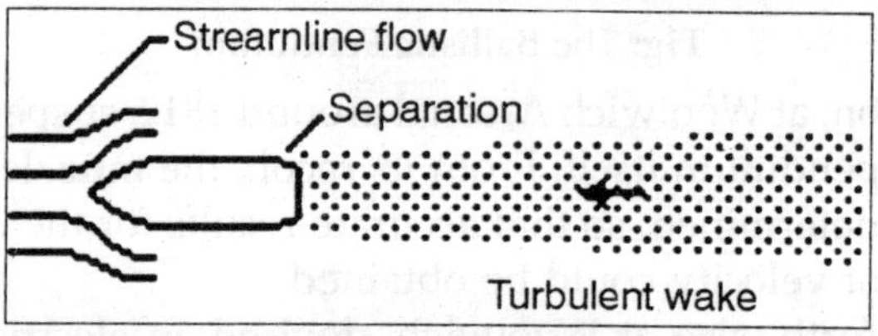

Fig: Subsonic Projectile

There is a stagnation point exactly at the nose of the projectile, whatever its shape, and otherwise the air flows smoothly around the obstruction. If there were no viscosity, the streamlines would close in again behind, and the air would exert no net force on the projectile: there would be no drag! Instead, flow in the diverging region is unstable, and a separation surface between streamline flow and the turbulent wake appears.

The air in the wake has been accelerated in the direction of the motion, representing a loss of kinetic energy of the projectile, and this movement is dissipated in turbulent eddies.

The pressure is low at the tail of the projectile, and this is the major contribution to the drag. A body that is not spherical tends to rotate so that its wake is as large in cross-section as possible, and the drag a maximum. This can be seen by dropping light objects, such as a leaf.

At around the speed of sound, Mach 1 or about 350 m/s, the drag increases abruptly, doubling or more, as we enter the transonic region, and then usually decreases slightly as the velocity is increased further. The reason for this change is a radical change in the flow around the projectile as M = 1 is passed.

The new feature is the appearance of shock-wave cones at the nose and tail, making an angle Ω with the trajectory. The Mach number M = 1/sin Ω. These are sudden jumps in pressure, carrying considerable energy away, which explains the increased drag.

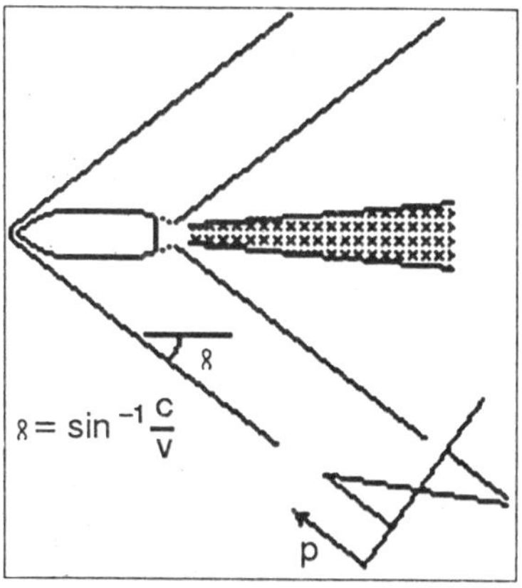

Fig: Supersonic Projectile

The tail drag from the turbulent wake still exists, but it is dominated by the shock-wave dissipation. The pressure signature of the shock waves is a kind of N, which is heard as a distinct crack that appears to originate some distance behind the projectile.

A strong shock wave travels at faster than the speed of sound, because of nonlinear propagation.

The curve of the shock wave around the nose of the projectile is accompanied by a high pressure, which appears as drag. The idea is to create as weak a shock wave as possible to reduce drag, and this is best done with a finely pointed nose.

Although aircraft must travel at faster than 350 m/s to enter the supersonic regime, wave speeds on the surface of water are small enough that ships are strongly affected by wave drag. The bow wave is analogous to the shock wave of a projectile, and carries away energy that must be replaced by the propulsion system.

There is also a turbulent wake, that sucks at the stern of the ship. The bows of the ship are pointed to reduce wave resistance. A canoe, which moves at less than the wave velocity is slender at each end to encourage streamline flow.

Although there are obvious analogies, there are essential differences, such as the two-dimensional nature of waves on water and their dispersive propagation. The kinematic viscosity of water is only 0.01 cm^2/s, a factor of 10 less than that of air.

The streamlining of subsonic vehicles is a different matter, which should concentrate on encouraging streamline flow as far as possible, and reducing the wake to as small a cross-section as possible. The shape of the front is of little consequence, so long as it does not generate turbulence. Almost any form will create an invisible region of dead air ahead of it that acts as smooth fairing. The rounded fairing seen on the fronts of road trailers is almost useless in reducing drag; it would have a greater effect on the rear, where it might discourage separation.

Any projections into the air stream create turbulence and consequently the drag. At the speed of automobiles, streamlining is only an esthetic matter.

Indeed many of the 'fast' cars actually have shapes that increase drag, and appurtenances that are applied in a largely useless fashion. Any smooth outline will do as well as any other. On railways, the resistance to motion from other causes is so small that air resistance makes a considerable, even a major contribution, above 50 mph, and streamlining can be economical, a fact often overlooked. As in many other cases, additional power is more cheaply applied than streamlining. If you want a fast car, a big engine will do the job better than excellent streamlining.

When $M > 1$, the undisturbed air is separated from air influenced by the projectile by a strong shock wave. There is one shock cone at the front, and another at the rear, both compressive. The turbulent wake exists as before. If the shock wave curves around the nose, high pressures result that are chiefly responsible for the increase in drag. By making the nose sharp and pointed, this force is reduced.

Making the tail slender has no effect, the turbulent wake being much as before. In contrast with subsonic streamlining, the body should now be slender in front, and blunt at the tail. This shows up in the shapes of subsonic and supersonic aircraft, where it is a critical matter to reduce drag as far as possible in either case.

The drag coefficient depends strongly on the angle between the axis and the direction of motion, called the angle of *yaw*. The drag coefficient also depends on the size of the projectile. It is greater per unit area for small cross sections than for large.

A projectile has an axis of symmetry, and the gun generally projects it so that this axis is parallel to the velocity as the projectile leaves the muzzle. This situation does not long remain, however, as a nonrotating projectile tumbles head over heels, greatly increasing the drag, and randomly deflecting the projectile to one side or another. Therefore, all projectiles were made spherical, the one shape that cannot tumble. There is, therefore, a very good reason why smoothbore guns fired only spherical bullets.

An alternative is to provide fins or feathers at the back of the projectile to force alignment by aerodynamic forces.

It was known as early as 1500 that imparting a spin along the axis stabilized a projectile, and prevented it from tumbling. This was done by cutting spiral grooves in the barrel, or *rifling*. These grooves became fouled so rapidly that the advantages of rifling were doubtful. In addition, the bullet had to fit the barrel tightly, and this was also a drawback, especially when the gun was loaded from the muzzle. Although rifled hand weapons were successfully used by 1800, their general adoption had to wait for the development of non-fouling smokeless powder.

The adoption of rifling meant that projectiles could become lengthened, achieving a larger projectile for the same diametre of barrel or *calibre*. The increase to supersonic muzzle velocities favoured the sharp-nosed shape with only a slight contraction at the tail, the *boat-shaped* projectile.

The nose of a projectile is usually *ogival,* that is, formed by rotating a circular arc tangent to the side of the projectile about the axis. The radius is specified in calibres (multiples of d). The nose may be rounded off by a spherical surface, also specified in calibres. When attempts were made to streamline the tail, it was found that such bullets tumbled and were useless. The shape of the bullet reflects its aerodynamics, not artistic design.

When the gun is considerably elevated, to achieve great range, or for the projectile to drop from above, the curved trajectory means that the axis of a spinning projectile cannot remain tangent to the trajectory, and complicated gyroscopic and drag effects enter. One result is *drift,* the departure of the trajectory from a plane, so that the projectile is deviated to right or left.

Golfers, and others, are well aware of this phenomenon that accompanies a spinning ball. Mortar rounds, which are always used in high trajectories, have fins and do not rotate.

For the accurate computation of a trajectory, the forces on the projectile as a function of yaw angle must be known, and the projectile treated as a rotating rigid body.

Then, with the drag known, the trajectory can be found by numerical integration, beginning with the initial velocity and spin. Not only is the problem a difficult and involved one, but the calculations are exceedingly tedious. In fact, they were so tedious that they stimulated the creation of digital computers in the 1940's, leading to the general-purpose computer, which has now revolutionized technology and life.

Chapter 6

Conservation of Energy

The conservation of energy is undoubtedly the single most important idea in physics. Strangely enough, although the basic idea of energy conservation was familiar to scientists from the time of Newton onwards, this crucial concept only moved to centre-stage in physics in about 1850 (i.e., when scientists first realised that heat was a form of energy).

According to the ideas of modern physics, energy is the substance from which all things in the Universe are made up. Energy can take many different forms: e.g., potential energy, kinetic energy, electrical energy, thermal energy, chemi- cal energy, nuclear energy, etc. In fact, everything that we observe in the world around us represents one of the multitudinous manifestations of energy.

Now, there exist processes in the Universe which transform energy from one form into another: e.g., mechanical processes (which are the focus of this course), thermal processes, electrical processes, nuclear processes, etc. However, all of these pro- cesses leave the total amount of energy in the Universe invariant. In other words, whenever, and however, energy is transformed from one form into another, it is always conserved. For a closed system (i.e., a system which does not exchange energy with the rest of the Universe), the above law of universal energy conserva-tion implies that the total energy of the system in question must remain constant in time.

ENERGY CONSERVATION DURING FREE-FALL

Consider a mass m which is falling vertically under the influence of gravity. We already know how to analyse the motion of such a mass. Let us employ this knowledge to search for an expression for the conserved energy during this process. The physics of free-fall under gravity is summarized by the three equations.

Let us examine the last of these equations:

$$v^2 = v^2_0 - 2\,g\,s.$$

Suppose that the mass falls from height h_1 to h_2, its initial velocity is v_1, and its final velocity is V_2. It follows that the net vertical displacement of the mass is $s = h_2 - h_1$. Moreover, $v_0 = v_1$ and $v = v_2$. Hence, the previous expression can be rearranged to give,

$$\frac{1}{2}mv_1^2 + mgh_1 = \frac{1}{2}mv_2^2 + mgh_2.$$

The above equation clearly represents a conservation law, of some description, since the left-hand side only contains quantities evaluated at the initial height, whereas the right-hand side only contains quantities evaluated at the final height. In order to clarify the meaning of Eq., let us define the kinetic energy of the mass,

$$K = \frac{1}{2}mv^2,$$

and the gravitational potential energy of the mass,

$$U = mgh.$$

Note that kinetic energy represents energy the mass possesses by virtue of its motion. Likewise, potential energy represents energy the mass possesses by virtue of its position. It follows that Eq. can be written,

$$E = K + U = \text{constant}.$$

Here, E is the total energy of the mass: i.e., the sum of its kinetic and potential energies. It is clear that E is a conserved quantity: i.e., although the kinetic and potential energies of the mass vary as it falls, its total energy remains the same.

Incidentally, the expressions and for kinetic and gravitational potential energy, respectively, are quite general, and do not just apply to free-fall under gravity. The mks unit of energy is called the joule (symbol J). In fact, 1 joule is equivalent to 1 kilogram metre-squared per second-squared, or 1 newtonmetre. Note that all forms of energy are measured in the same units (otherwise the idea of energy conservation would make no sense).

One of the most important lessons which students learn during their studies is that there are generally many different paths to the same result in physics. Now, we have already analysed free-fall under gravity using Newton's laws of motion.

However, it is illuminating to re-examine this problem from the point of view of energy conservation. Suppose that a mass m is dropped from rest and falls a distance h. What is the final velocity v of the mass? Well, according to Eq., if energy is conserved then,

$$\Delta K = -\Delta U:$$

i.e., any increase in the kinetic energy of the mass must be offset by a corresponding decrease in its potential energy. Now, the change in potential energy of the mass is simply $\Delta U = mg\, s = -mg\, h$, where $s = -h$ is its net vertical displacement. The change in kinetic energy is simply $\Delta K = (1/2)\, mv^2$, where v is the final velocity. This follows because the initial kinetic energy of the mass is zero (since it is initially at rest). Hence, the above expression yields,

$$\frac{1}{2}mv^2 = mgh,$$

or,

$$v = \sqrt{2gh}.$$

Suppose that the same mass is thrown upwards with initial velocity v. What is the maximum height h to which it rises? Well, it is clear from Eq. that as the mass rises its potential energy increases. It, therefore, follows from energy conservation that its kinetic energy must decrease with height. Note, however, from Eq.that kinetic energy can never be negative (since it is the product of the two positive definite quantities, m and $v^2/2$). Hence, once the mass has risen to a height h which is such that its kinetic energy is reduced to zero it can rise no further, and must, presumably, start to fall. The change in potential energy of the mass in moving from its initial height to its maximum height is m g h. The corresponding change in kinetic energy is – (1/2)mv^2; since (1/2)mv^2 is the initial kinetic energy, and the final kinetic energy is zero. It follows from Eq. that – (1/2)mv^2 = – $mg\,h$, which can be rearranged to give,

$$h = \frac{v^2}{2g}.$$

It should be noted that the idea of energy conservation—although extremely useful—is not a replacement for Newton's laws of motion. For instance, in the previous example, there is no way in which we can deduce how long it takes the mass to rise to its maximum height from energy conservation alone—this information can only come from the direct application of Newton's laws.

WORK

We have seen that when a mass free-falls under the influence of gravity some of its kinetic energy is transformed into potential energy, or vice versa. Let us now investigate, in detail, how this transformation is effected. The mass falls because it is subject to a downwards gravitational force of magnitude m g. It stands to reason, therefore, that the transformation of kinetic into potential energy is a direct consequence of the action of this force.

This is, perhaps, an appropriate point at which to note that the concept of gravitational potential energy—although extremely useful—is, strictly speaking, fictitious. To be more exact, the potential energy of a body is not an intrinsic property of that body (unlike its kinetic energy). In fact, the gravitational potential energy of a given body is stored in the gravitational field which surrounds it. Thus, when the body rises, and its potential energy consequently increases by an amount ΔU; in reality, it is the energy of the gravitational field surrounding the body which increases by this amount. Of course, the increase in energy of the gravitational field is offset by a corresponding decrease in the body's kinetic energy. Thus, when we speak of a body's kinetic energy being transformed into potential energy, we are really talking about a flow of energy from the body to the surrounding gravitational field. This energy flow is mediated by the gravitational force exerted by the field on the body in question.

Incidentally, according to Einstein's general theory of relativity (1917), the gravitational field of a mass consists of the local distortion that mass induces in the fabric of space-time. Fortunately, however, we do not need to understand general relativity in order to talk about gravitational fields or gravitational po-tential energy. All we need to know is that a gravitational field stores energy without loss: i.e., if a given mass rises a certain distance, and, thereby, gives up a certain amount of energy to the surrounding gravitational field, then that field will return this energy to the mass—without loss—if the mass falls by the same distance. In physics, we term such a field a conservative field.

Suppose that a mass m falls a distance h. During this process, the energy of the gravitational field decreases by a certain amount (i.e., the fictitious potential energy of the mass decreases by a certain amount), and the body's kinetic energy increases by a corresponding amount. This transfer of energy, from the field to the mass, is, presumably, mediated by the gravitational force -m g (the minus sign indicates that the force is directed downwards) acting on the mass. In fact, given that $U = mg\, h$, it follows from Eq. that

$$\Delta K = f\Delta h.$$

In other words, the amount of energy transferred to the mass (i.e., the increase in the mass's kinetic energy) is equal to the product of the force acting on the mass and the distance moved by the mass in the direction of that force.

In physics, we generally refer to the amount of energy transferred to a body, when a force acts upon it, as the amount of work W performed by that force on the body in question. It follows from Eq. that when a gravitational force f acts on a body, causing it to displace a distance x in the direction of that force, then the net work done on the body is,

$$W = fx.$$

It turns out that this equation is quite general, and does not just apply to gravitational forces. If W is positive then energy is transferred to the body, and its intrinsic energy consequently increases by an amount *W*. This situation occurs whenever a body moves in the same direction as the force acting upon it. Likewise, if *W* is negative then energy is transferred from the body, and its intrinsic energy consequently decreases by an amount $|W|$. This situation occurs whenever a body moves in the opposite direction to the force acting upon it. Since an amount of work is equivalent to a transfer of energy, the mks unit of work is the same as the mks unit of energy: namely, the joule.

We have made two assumptions which are not universally valid. Firstly, we have assumed that the motion of the body upon which the force acts is both 1-dimensional and parallel to the line of action of the force. Secondly, we have assumed that the force does not vary with position. Let us attempt to relax these two assumptions, so as to obtain an expression for the work W done by a general force f.

Let us start by relaxing the first assumption. Suppose, for the sake of argument, that we have a mass m which moves under gravity in 2-dimensions. Let us adopt the coordinate system shown in Fig. 35, with z representing vertical distance, and *x* representing horizontal distance. The vector acceleration of the mass is simply $a = (0, -g)$. Here, we are neglecting the redundant y-component, for the sake of simplicity. The physics of motion under gravity in more than 1-dimension is summarized by the three equations -.

Let us examine the last of these equations:

$$v^2 = v^2_0 + 2a{\cdot}s.$$

Here, v_0 is the speed at $t = 0$, v is the speed at $t = t$, and $s = (\Delta x, \Delta z)$ is the net displacement of the mass during this time interval. Recalling the definition of a scalar product [*i.e.*, $a{\cdot}b = (a_x\, b_x + a_y\, b_y + a_z\, b_z)$], the above equation can be rearranged to give,

$$\frac{1}{2}mv^2 - \frac{1}{2}mv_0^2 = -mg\Delta z.$$

Since the right-hand side of the above expression is manifestly the increase in the kinetic energy of the mass between times t = 0 and t = t, the left-hand side must equal the decrease in the mass's potential energy during the same time interval. Hence, we arrive at the following expression for the gravitational potential energy of the mass:

$$U = mgz.$$

Of course, this expression is entirely equivalent to our previous expression for gravitational potential energy, Eq.. The above expression merely makes manifest a point which should have been obvious anyway: namely, that the gravitational potential energy of a mass only depends on its height above the ground, and is quite independent of its horizontal displacement.

Let us now try to relate the flow of energy between the gravitational field and the mass to the action of the gravitational force, $f = (0, -mg)$.

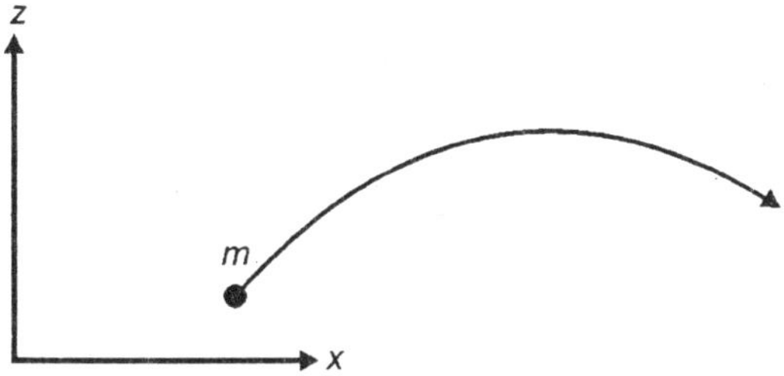

Fig. Coordinate System for 2-dimensional Motion Under Gravity

Can be rewritten,

$$\Delta K = W = f \cdot s.$$

In other words, the work *W* done by the force f is equal to the scalar product of f and the vector displacement s of the body upon which the force acts. It turns out that this result is quite general, and does not just apply to gravitational forces.

Figure 36 is a visualization of the definition. The work *W* performed by a force *f* when the object upon which it acts is subject to a displacement *s* is,

$$W = |f|\ |s| \cos\theta.$$

where θ is the angle subtended between the directions of f and s. In other words, the work performed is the product of the magnitude of the force, $|f|$, and the displacement of the object in the direction of that force, $|s| \cos\theta$. It follows that any component of the displacement in a direction perpendicular to the force generates zero work. Moreover, if the displacement is entirely perpendicular to the direction of the force (*i.e.*, if $\theta = 90^\circ$) then no work is performed, irrespective of the nature of the force. As before, if the displacement has a component in the same direction as the force (i.e., if $\theta < 90^\circ$) then positive work is performed Likewise, if the displacement has a component in the opposite direction to the force (*i.e.*, if $\theta > 90^\circ$) then negative work is performed.

Suppose, now, that an object is subject to a force f which varies with position. What is the total work done by the force when the object moves along some general trajectory in space between points A and B (say)? Well, one way in which we could approach this problem would be to approximate the trajectory as a series of N straight-line segments, as shown in Fig. Suppose

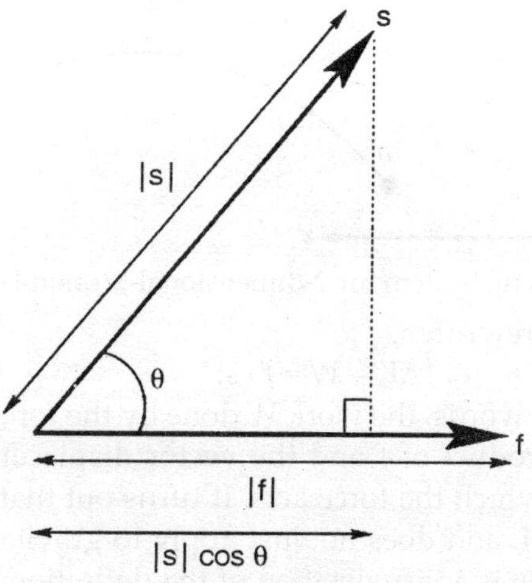

Fig. Definition of work

That the vector displacement of the ith segment is Δr_i. Suppose, further, that N is sufficiently large that the force f does not vary much along each segment. In fact, let the average force along the ith segment be f_i. We shall assume that formula —which is valid for constant forces and straight-line displacements—holds good for each segment. It follows that the net work done on the body, as it moves from point A to point B, is approximately,

$$W \simeq \sum_{i=1}^{N} f_i \cdot \Delta r_i.$$

We can always improve the level of our approximation by increasing the number Nof the straight-line segments which we use to approximate the body's trajectory between points A and B. In fact, if we take the limit N $\to \infty$ then the above expression becomes exact:

$$W = \lim_{N \to \infty} \sum_{i=1}^{N} f_i \cdot \Delta r_i = \int_A^B f(r) \cdot dr.$$

and the mathematical construct R_{BA} $f(r)$ _dr is termed a line-integral.

The meaning of Eq. becomes a lot clearer if we restrict our attention to 1-dimensional motion. Suppose, therefore, that an object moves in 1-dimension, with displacement x, and is subject to a varying force $f(x)$ (directed along the x-axis). What is the work done by this force when the object moves from x_A

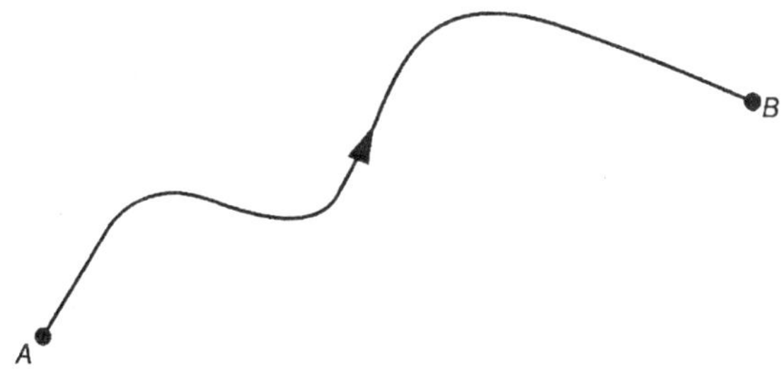

Fig. Possible Trajectory of an Object in a Variable force- field

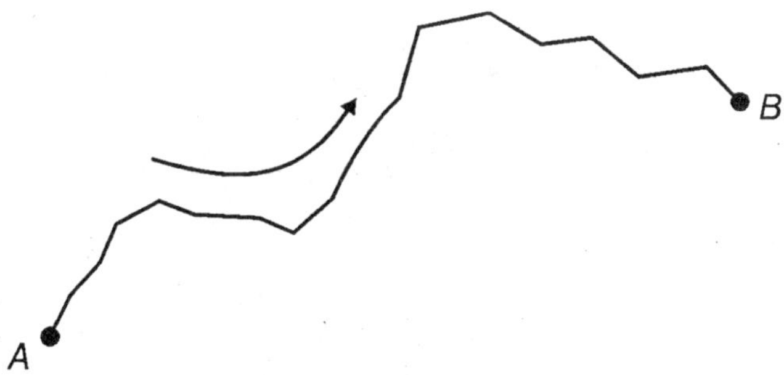

Fig. Approximation to the Previous Trajectory using Straight-line Segments

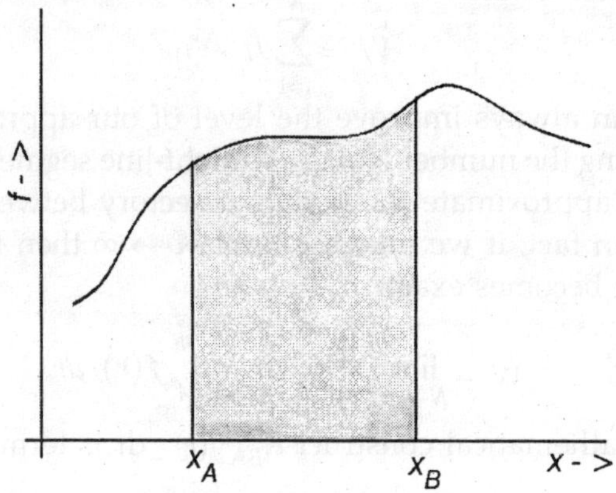

Fig. Work Performed by a 1-dimensional Force to x_B Well, a Straight-forward Application of Eq. [with $f = (f, 0, 0)$ and $dr = (dx, 0, 0)$] Yields.

$$W = \int_{x_A}^{x_B} f(x)dx.$$

In other words, the net work done by the force as the object moves from displacement x_A to x_B is simply the area under the $f(x)$ curve between these two points.

Let us, finally, round-off this discussion by re-deriving the so-called workenergy theorem, Eq., in 1-dimension, allowing for a non-constant force. According to Newton's second law of motion,

$$f = m\frac{d^2x}{dt^2}.$$

Combining Eqs. and, we obtain,

$$W = \int_{x_A}^{x_B} m\frac{d^2x}{dt^2}dx = \int_{t_A}^{t_B} m\frac{d^2x\,dx}{dt^2\,dt}dt$$

$$= \int_{t_A}^{t_B} \frac{d}{dt}\left[\frac{m}{2}\left(\frac{dx}{dt}\right)^2\right]dt,$$

where,

$$x(t_A) = x_A \text{ and } x(t_B) = x_B.$$

It follows that,

$$W = \frac{1}{2}mv_A^2 - \frac{1}{2}mv_A^2 = \Delta K,$$

where,

$$v_A = (dx = dt)t_A \text{ and } v_B = (dx = dt)t_B.$$

Thus, the net work performed on a body by a non-uniform force, as it moves from point A to point B, is equal to the net increase in that body's kinetic energy between these two points. This result is completely general, and does not just apply to 1-dimensional motion.

Suppose, finally, that an object is subject to more than one force. How do we calculate the net work W performed by all these forces as the object moves from point A to point B? One approach would be to calculate the work done by each force, taken in isolation, and then to sum the results. In other words, defining,

$$W_i = \int_A^B f_i(r) \cdot dr$$

as the work done by the ith force, the net work is given by

$$W = \sum_i W_i.$$

An alternative approach would be to take the vector sum of all the forces to find the resultant force,

$$f = \sum_i f_i,$$

and then to calculate the work done by the resultant force:

$$W = \int_A^B f(r) \cdot dr.$$

It should, hopefully, be clear that these two approaches are entirely equivalent.

CONSERVATIVE AND NON-CONSERVATIVE FORCE- FIELDS

Suppose that a non-uniform force- field $f(r)$ acts upon an object which moves along a curved trajectory, labeled path 1, from point A to point B. As we have seen, the work W_1 performed by the force- field on the object can be written as a line-integral along this trajectory:

$$W_1 = \int_{A \to B:\text{path 1}} f \cdot dr.$$

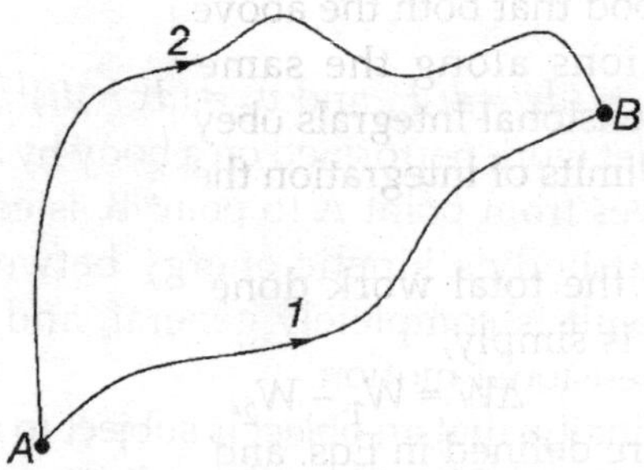

Fig. Two Alternative paths Between Points A and B

Suppose that the same object moves along a different trajectory, labeled path 2, between the same two points. In this case, the work W_2 performed by the force- field is,

$$W_2 = \int_{A \to B:\text{path 2}} f \cdot dr.$$

Basically, there are two possibilities. Firstly, the line-integrals and might depend on the end points, A and B, but not on the path taken between them, in which case $W_1 = W_2$. Secondly, the line-integrals and might depend both on the end points, A and B, and the path taken between them, in which case $W_1 \neq W_2$ (in general). The first possibility corresponds to what physicists term a conservative force- field, whereas the second possibility corresponds to a non-conservative force-field.

What is the physical distinction between a conservative and a non-conservative force- field? Well, the easiest way of answering this question is to slightly modify the problem discussed above. Suppose, now, that the object moves from point A to point B along path 1, and then from point B back to point A along path 2. What is the total work done on the object by the force- field as it executes this closed circuit? Incidentally, one fact which should be clear from the definition of a line-integral is that if we simply reverse the path of a given integral then the value of that integral picks up a minus sign: in other words,

$$\int_A^B f \cdot dr = -\int_B^A f \cdot dr,$$

where it is understood that both the above integrals are taken in opposite directions along the same path. Recall that conventional 1-dimensional integrals obey an analogous rule: i.e., if we swap the limits of integration then the integral picks up a minus sign.

It follows that the total work done on the object as it executes the circuit is simply,

$$\Delta W = W_1 - W_2,$$

where W_1 and W_2 are defined in Eqs. and, respectively. There is a minus sign in front of W_2 because we are moving from point B to point A, instead of the other way around. For the case of a conservative field, we have $W_1 = W_2$.

Hence, we conclude that,

$$\Delta W = 0.$$

In other words, the net work done by a conservative field on an object taken around a closed loop is zero. This is just another way of saying that a conservative field stores energy without loss: i.e., if an object gives up a certain amount of energy to a conservative field in traveling from point A to point B, then the field returns this energy to the object—without loss—when it travels back to point B. For the case of a non-conservative field, $W_1 \neq W_2$.

Hence, we conclude that,

$$\Delta W \neq 0.$$

In other words, the net work done by a non-conservative field on an object taken around a closed loop is non-zero. In practice, the net work is invariably negative. This is just another way of saying that a non-conservative field dissipates energy: i.e., if an object gives up a certain amount of energy to a non-conservative field in traveling from point A to point B, then the field only returns part, or, perhaps, none, of this energy to the object when it travels back to point B. The remainder is usually dissipated as heat.

What are typical examples of conservative and non-conservative fields? Well, a gravitational field is probably the most well-known example of a conservative field. A typical example of a non-conservative field might consist of an object moving over a rough horizontal surface. Suppose, for the sake of simplicity, that the object executes a closed circuit on the surface which is made up entirely of straight-line segments.

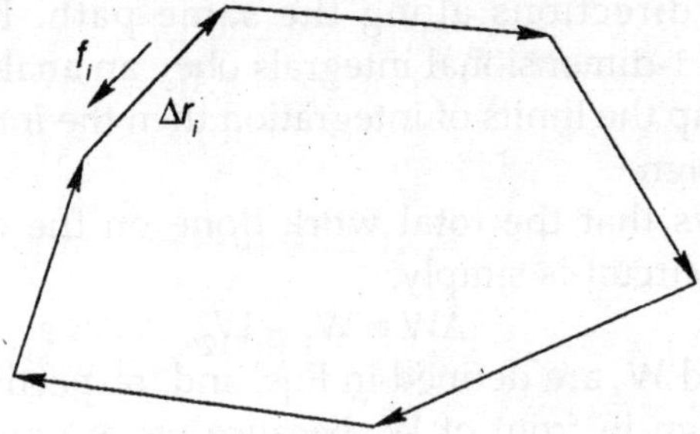

Fig. Closed circuit over a Rough Horizontal Surface

Let Δr_i represent the vector displacement of the ith leg of this circuit. Suppose that the frictional force acting on the object as it executes this leg is f_i. One thing that we know about a frictional force is that it is always directed in the opposite direction to the instantaneous direction of motion of the object upon which it acts.

Hence, $f_i \propto -\Delta r_i$. It follows that $f_i \cdot \Delta r_i = -|f_i|\ |\Delta r_i|$. Thus, the net work performed by the frictional force on the object, as it executes the circuit, is given by,

$$\Delta W = \sum_i f_i \cdot \Delta r_i = -\sum_i |f_i||\Delta r_i| < 0.$$

The fact that the net work is negative indicates that the frictional force continually drains energy from the object as it moves over the surface. This energy is actu-ally dissipated as heat (we all know that if we rub two rough surfaces together, sufficiently vigourously, then they will eventually heat up: this is how mankind first made fire) and is, therefore, lost to the system. Hence, friction is an example of a non-conservative force, because it dissipates energy rather than storing it.

POTENTIAL ENERGY

Consider a body moving in a conservative force- field $f(r)$. Let us arbitrarily pick some point O in this field. We can define a function $U(r)$ which possesses a unique value at every point in the field. The value of this function associated with some general point R is simply,

$$U(R) = -\int_O^R f \cdot dr.$$

In other words, $U(R)$ is just the energy transferred to the field (i.e., minus the work done by the field) when the body moves from point O to point R. Of course, the value of U at point O is zero: i.e., $U(O) = 0$.

Note that the above definition uniquely specifies $U(R)$, since the work done when a body moves between two points in a conservative force- field is independent of the path taken between these points.

Furthermore, the above definition would make no sense in a non-conservative field, since the work done when a body moves between two points in such a field is dependent on the chosen path: hence, $U(R)$ would have an infinite number of different values corresponding to the infinite number of different paths the body could take between points O and R.

According to the work-energy theorem,

$$\Delta K = \int_O^R f \cdot dr.$$

In other words, the net change in the kinetic energy of the body, as it moves from point O to point R, is equal to the work done on the body by the force- field during this process. However, comparing with Eq., we can see that,

$$\Delta K = U(O) - U(R) = -\Delta U.$$

In other words, the increase in the kinetic energy of the body, as it moves from point O to point R, is equal to the decrease in the function U evaluated between these same two points.

Another way of putting this is,

$$E = K + U = \text{constant}:$$

i.e., the sum of the kinetic energy and the function U remains constant as the body moves around in the force- field. It should be clear, by now, that the function U represents some form of potential energy.

The above discussion leads to the following important conclusions. Firstly, it should be possible to associate a potential energy (i.e., an energy a body pos-sesses by virtue of its position) with any conservative force- field. Secondly, any force- field for which we can define a potential energy must necessarily be con-servative. For instance, the existence of gravitational potential energy is proof that gravitational fields are conservative.

Thirdly, the concept of potential energy is meaningless in a non-conservative force- field (since the potential energy at a given point cannot be uniquely defined).

Fourthly, potential energy is only defined to within an arbitrary additive constant. In other words, the point in space at which we set the potential energy to zero can be chosen at will. This im-plies that only differences in potential energies between different points in space have any physical significance.

For instance, we have seen that the definition of gravitational potential energy is $U = \mathrm{m}\, gz$, where z represents height above the ground. However, we could just as well write $U = mg\,(z - z_0)$, where z_0 is the height of some arbitrarily chosen reference point (e.g., the top of Mount Ever-est, or 0the bottom of the Dead Sea).

Fifthly, the difference in potential energy between two points represents the net energy transferred to the associated force-field when a body moves between these two points. In other words, potential energy is not, strictly speaking, a property of the body—instead, it is a property of the force-field within which the body moves.

HOOKE'S LAW

Consider a mass m which slides over a horizontal frictionless surface. Suppose that the mass is attached to a light horizontal spring whose other end is anchored to an immovable object. Let x be the extension of the spring: *i.e.*, the difference between the spring's actual length and its unstretched length.

Ob-viously, x can also be used as a coordinate to determine the horizontal displace-ment of the mass. According to Hooke's law, the force f that the spring exerts on the mass is directly proportional to its extension, and always acts to reduce this extension.

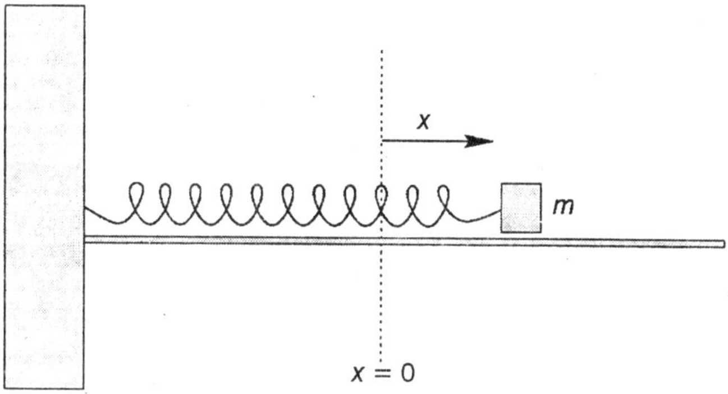

Fig. Mass on a spring

Hence, we can write,

$$f = -kx,$$

where the positive quantity k is called the force constant of the spring. Note that the minus sign in the above equation ensures that the force always acts to reduce the spring's extension: e.g., if the extension is positive then the force acts to the left, so as to shorten the spring.

According to Eq., the work performed by the spring force on the mass as it moves from displacement x_A to x_B is,

$$W = \int_{x_A}^{x_B} f(x)dx = -k\int_{z_A}^{x_B} x\,dx = -\left[\frac{1}{2}k\,x_B^2 - \frac{1}{2}k\,x_A^2\right].$$

Note that the right-hand side of the above expression consists of the difference between two factors: the first only depends on the final state of the mass, whereas the second only depends on its initial state. This is a sure sign that it is possible to associate a potential energy with the spring force. Equation, which is the basic definition of potential energy, yields

$$U(x_B) - U(x_A) = -\int_{x_A}^{x_B} f(x)dx = \frac{1}{2}k\,x_B^2 - \frac{1}{2}k\,x_A^2.$$

Hence, the potential energy of the mass takes the form,

$$U(x) = \frac{1}{2}k\,x^2.$$

Note that the above potential energy actually represents

energy stored by the spring—in the form of mechanical stresses—when it is either stretched or compressed. Incidentally, this energy must be stored without loss, otherwise the concept of potential energy would be meaningless. It follows that the spring force is another example of a conservative force.

It is reasonable to suppose that the form of the spring potential energy is somehow related to the form of the spring force. Let us now explicitly investigate this relationship. If we let $x_B \rightarrow x$ and $x_A \rightarrow 0$ then Eq. gives

$$U(x) = \int_0^x f(x')dx'.$$

We can differentiate this expression to obtain,

$$f = \frac{d\bar{U}}{dx}.$$

Thus, in 1-dimension, a conservative force is equal to minus the derivative (with respect to displacement) of its associated potential energy. This is a quite general result. For the case of a spring force: $U = (1/2)\, kx^2$, so $f = -\, dU/dx = -kx$.

As is easily demonstrated, the 3-dimensional equivalent to Eq. is,

$$f = \left(\frac{\partial U}{\partial x}, \frac{\partial U}{\partial y}, \frac{\partial U}{\partial z} \right).$$

For example, we have seen that the gravitational potential energy of a mass m moving above the Earth's surface is $U = mg\, z$, where z measures height off the ground. It follows that the associated gravitational force is,

$$f = (0, 0, -\, mg).$$

In other words, the force is of magnitude mg, and is directed vertically downward. The total energy of the mass shown is the sum of its kinetic and potential energies:

$$E = K + U = K + \frac{1}{2}kx^2.$$

Of course, E remains constant during the mass's motion. Hence, the above ex- pression can be rearranged to give

$$K = E - \frac{1}{2}kx^2.$$

Since it is impossible for a kinetic energy to be negative, the above expression suggests that $|x|$ can never exceed the value,

$$x_0 = \sqrt{\frac{2E}{k}}.$$

Here, x_0 is termed the amplitude of the mass's motion. Note that when x attains its maximum value x_0, or its minimum value $-x_0$, the kinetic energy is momentarily zero (*i.e.*, $K = 0$).

MOTION IN A GENERAL 1-DIMENSIONAL POTENTIAL

Suppose that the curve $U(x)$ in Fig. 43 represents the potential energy of some mass m moving in a 1-dimensional conservative force- field. For instance, $U(x)$ might represent the gravitational potential energy of a cyclist freewheeling in a hilly region. Note that we have set the potential energy at infinity to zero. This is a useful, and quite common, convention (recall that potential energy is undefined to within an arbitrary additive constant). What can we deduce about the motion of the mass in this potential? Well, we know that the total energy, E—which is the sum of the kinetic energy, K, and the potential energy, U—is a constant of the motion.

Hence, we can write,

$$K(x) = E - U(x).$$

Now, we also know that a kinetic energy can never be negative, so the above expression tells us that the motion of the mass is restricted to the region (orregions) in which the potential energy curve $U(x)$ falls below the value E. Suppose that the total energy of the system is E_0.

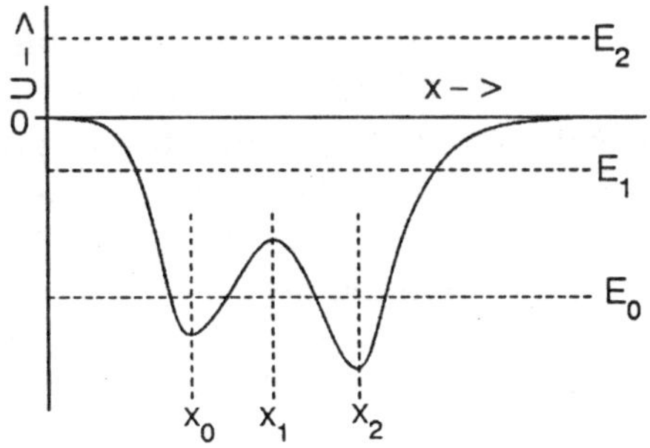

Fig. General 1-dimensional Potential

It is clear, from the figure, that the mass is trapped inside one or other of the two dips in the potential—these dips are generally referred to as potential wells. Suppose that we now raise the energy to E_1. In this case, the mass is free to enter or leave each of the potential wells, but its motion is still bounded to some extent, since it clearly cannot move off to inflnity. Finally, let us raise the energy to E_2.

Now the mass is unbounded: i.e., it can move off to infinity. In systems in which it makes sense to adopt the convention that the potential energy at infinity is zero, bounded systems are characterized by $E < 0$, whereas unbounded systems are characterized by $E > 0$.

The above discussion suggests that the motion of a mass moving in a potential generally becomes less bounded as the total energy E of the system increases. Conversely, we would expect the motion to become more bounded as E decreases. In fact, if the energy becomes sufficiently small, it appears likely that the system will settle down in some equilibrium state in which the mass is stationary. Let us try to identify any prospective equilibrium states in Fig. 43. If the mass remains stationary then it must be subject to zero force (otherwise it would accelerate). Hence, according to Eq., an equilibrium state is characterized by,

$$\frac{dU}{dx} = 0.$$

In other words, a equilibrium state corresponds to either a maximum or a minimum of the potential energy curve $U(x)$. It can be seen that the U(x) curve shown has three associated equilibrium states: these are located at $x = x_0$, $x = x_1$, and $x = x_2$. Let us now make a distinction between stable equilibrium points and unstable equilibrium points. When the system is slightly perturbed from a stable equilibrium point then the resultant force f should always be such as to attempt to return the system to this point. In other words, if $x = x_0$ is an equilibrium point, then we require,

$$\left.\frac{df}{dx}\right|_{x_0} < 0$$

for stability: i.e., if the system is perturbed to the right, so that $x - x_0 > 0$, then the force must act to the left, so that $f < 0$, and vice versa. Likewise, if

$$\left.\frac{df}{dx}\right|_{x_0} > 0$$

then the equilibrium point $x = x_0$ is unstable. It follows, from Eq., that stable equilibrium points are characterized by

$$\frac{d^2U}{dx^2} > 0.$$

In other words, a stable equilibrium point corresponds to a minimum of the po-tential energy curve $U(x)$. Likewise, an unstable equilibrium point corresponds to a maximum of the $U(x)$ curve. Hence, we conclude that $x = x_0$ and $x = x_2$ are stable equilibrium points, in Fig. 43, whereas $x = x_1$ is an unstable equilibrium point. Of course, this makes perfect sense if we think of $U(x)$ as a gravitational potential energy curve, in which case U is directly proportional to height. All we are saying is that it is easy to confine a low energy mass at the bottom of a valley, but very difficult to balance the same mass on the top of a hill (since any slight perturbation to the mass will cause it to fall down the hill). Note, finally, that if

$$\frac{dU}{dx} = \frac{d^2U}{dx^2} = 0$$

at any point (or in any region) then we have what is known as a neutral equilib-rium point. We can move the mass slightly off such a point and it will still remain in equilibrium (i.e., it will neither attempt to return to its initial state, nor will it continue to move). A neutral equilibrium point corresponds to a flat spot in a U(x) curve.

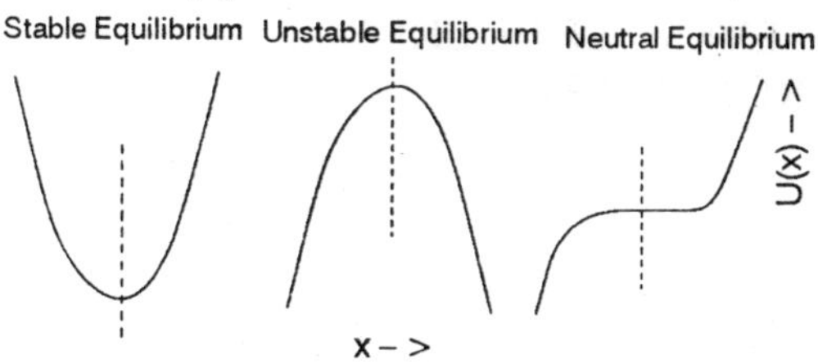

Fig. Different types of equilibrium

POWER

Suppose that an object moves in a general force- field $f(r)$. We now know how to calculate how much energy flows from the force- field to the object as it moves along a given path between two points. Let us now consider the rate at which this energy flows. If dW is the amount of work that the force- field performs on the mass in a time interval dt then the rate of working is given by,

$$P = \frac{dW}{dt}.$$

In other words, the rate of working—which is usually referred to as the power—is simply the time derivative of the work performed. Incidentally, the mks unit of power is called the watt (symbol W). In fact, 1 watt equals 1 kilogram metresquared per second-cubed, or 1 joule per second.

Suppose that the object displaces by dr in the time interval dt. By definition, the amount of work done on the object during this time interval is given by,

$$dW = f \cdot dr.$$

It follows from Eq. that,

$$P = f \cdot v,$$

where $v = dr/dt$ is the object's instantaneous velocity. Note that power can be positive or negative, depending on the relative directions of the vectors f and v. If these two vectors are mutually perpendicular then the power is zero. For the case of 1-dimensional motion, the above expression reduces to,

$$P = fv.$$

In other words, in 1-dimension, power simply equals force times velocity.

WORKED EXAMPLE BUCKET LIFTED FROM A WELL

Question: A man lifts a 30 kg bucket from a well whose depth is 150 m. Assuming that the man lifts the bucket at a constant rate, how much work does he perform?

Answer: Let m be the mass of the bucket and h the depth of the well. The gravitational force f acting on the bucket is of magnitude mg and is directed vertically downwards. Hence, $f' = -mg$ (where upward is defined to be positive). The net upward displacement of the bucket is h. Hence, the work W′ performed by the gravitational force is the product of the (constant) force and the displacement of the bucket along the line of action of that force:

$$W' = f' h = -mgh.$$

Note that W′ is negative, which implies that the gravitational field surrounding the bucket gains energy as the bucket is lifted. In order to lift the bucket at a constant rate, the man must exert a force f on the bucket which balances (and very slightly exceeds) the force due to gravity. Hence, $f = -f'$. It follows that the work W done by the man is,

$$W = fh = mgh = 30 \times 150 \times 9.81 = 4.415 \times 10^4 \text{ J}.$$

Note that the work is positive, which implies that the man expends energy whilst lifting the bucket. Of course, since $W = -W'$, the energy expended by the man equals the energy gained by the gravitational field.

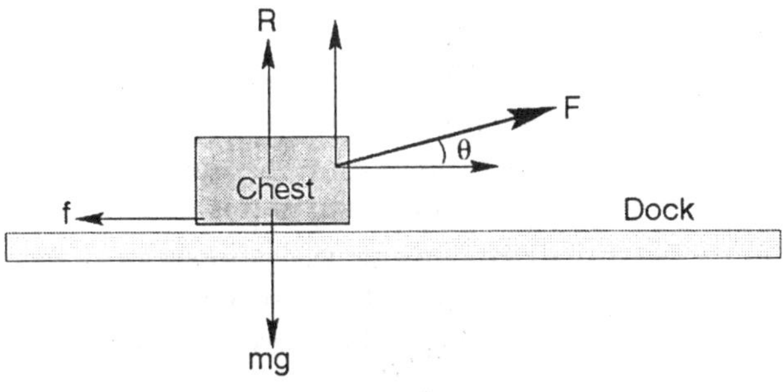

WORKED EXAMPLE DRAGGING A TREASURE CHEST

Question: A pirate drags a 50 kg treasure chest over the rough surface of a dock by exerting a constant force of 95 N acting at an angle of 15- above the horizon-tal. The chest moves 6 m in a straight line, and the coef_cient of kinetic friction between the chest and the dock is 0:15. How much work does the pirate perform? How much energy is dissipated as heat via friction? What is the final velocity of the chest?

Answer: Referring to the diagram, the force F exerted by the pirate can be re-solved into a horizontal component F cos θ and a vertical component $F \sin \theta$. Since the chest only moves horizontally, the vertical component of F performs zero work. The work W performed by the horizontal component is simply the magnitude of this component times the horizontal distance x moved by the chest:

$$W = F \cos\theta\, x = 95 \times \cos 15^\circ \times 6 = 550.6 \text{ J.}$$

The chest is subject to the following forces in the vertical direction: the down-ward force m g due to gravity, the upward reaction force R due to the dock, and the upward component $F \sin\theta$ of the force exerted by the pirate. Since the chest does not accelerate in the vertical direction, these forces must balance.

Hence,

$$R = mg - F \sin\theta = 50 \times 9.81 - 95 \times \sin 15^\circ = 465.9 \text{ N.}$$

The frictional force f is the product of the coef_cient of kinetic friction μ_k and the normal reaction R, so

$$f = \mu_k R = 0.15 \times 465.9 = 69.89 \text{ N.}$$

The work W′ done by the frictional force is,

$$W' = -fx = -69.89 \times 6 = -419.3 \text{ J.}$$

Note that there is a minus sign in front of the f because the displacement of the chest is in the opposite direction to the frictional force. The fact that W' is negative indicates a loss of energy by the chest: this energy is dissipated as heat via friction. Hence, the dissipated energy is 419:3 J.

The final kinetic energy K of the chest (assuming that it is initially at rest) is the difference between the work W done by the pirate and the energy $-W'$ dissipated as heat.

Hence,

$$K = W + W' = 550.6 - 419.3 = 131.3 \text{ J.}$$

Since $K = (1/2)\, mv^2$, the final velocity of the chest is,

$$v = \sqrt{\frac{2K}{m}} = \sqrt{\frac{2 \times 131.3}{50}} = 2.29 m/s.$$

WORKED EXAMPLE: STRETCHING A SPRING

Question: The force required to slowly stretch a spring varies from 0 N to 105 N as the spring is extended by 13 cm from its unstressed length. What is the force constant of the spring? What work is done in stretching the spring? Assume that the spring obeys Hooke's law.

Answer: The force f that the spring exerts on whatever is stretching it is $f = -kx$, where k is the force constant, and x is the extension of the spring. The minus sign indicates that the force acts in the opposite direction to the extension. Since the spring is stretched slowly, the force f' which must be exerted on it is (almost) equal and opposite to f. Hence, $f' = -f = kx$. We are told that $f' = 105$ N when $x = 0.13m$. It follows that,

$$k = \frac{105}{0.13} = 807.7 N/m.$$

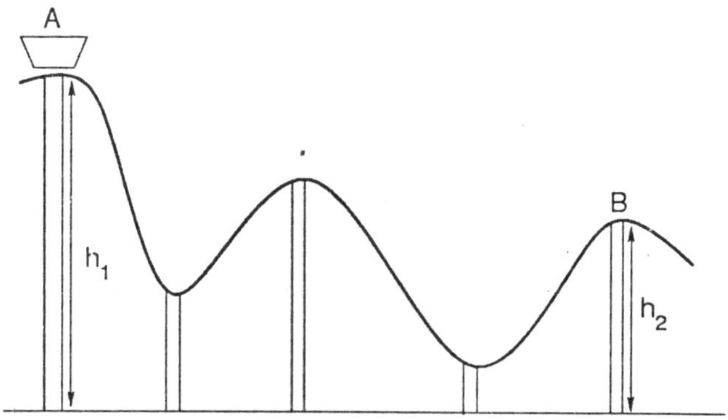

Fig. The work *W¢* **done by the** External Force in Extending the Spring from 0 to x is

$$W' = \int_0^x f'dx = k\int_0^x x\,dx = \frac{1}{2}kx^2.$$

Hence,

$$W' = 0.5 \times 807.7 \times 0.13^2 = 6.83 \text{ J}.$$

WORKED EXAMPLE: ROLLER COASTER RIDE

Question: A roller coaster cart of mass $m = 300$ kg starts at rest at point A, whose height off the ground is $h_1 = 25m$, and a little while later reaches point B, whose height off the ground is $h_2 = 7m$. What is the potential energy of the cart relative to the ground at point A? What is the speed of the cart at point B, neglecting the effect of friction?

Answer: The gravitational potential energy of the cart with respect to the ground at point A is,

$$U_A = mg\,h_1 = 300 \times 9.81 \times 25 = 7.36 \times 10^4 \text{ J.}$$

Likewise, the potential energy of the cart at point B is,

$$U_B = mg\,h_2 = 300 \times 9.81 \times 7 = 2.06 \times 10^4 \text{ J.}$$

Hence, the change in the cart's potential energy in moving from point A to point B is,

$$\Delta U = U_B - U_A = 2.06 \times 10^4 - 7.36 \times 10^4 = -5.30 \times 10^4 \text{ J.}$$

By energy conservation, $\Delta K = -\Delta U$, where K represents kinetic energy. However, since the initial kinetic energy is zero, the change in kinetic energy ΔK is equivalent to the final kinetic energy K_B.

Thus,

$$K_B = -\,\Delta U = 5.30 \times 10^4 \text{ J.}$$

Now, $K_B = (1/2)\, mv^2_B$, where v_B is the final speed.

Hence,

$$v_B = \sqrt{\frac{2K_B}{m}} = \sqrt{\frac{2 \times 5.30 \times 10^4}{300}} = 18.8 m/s.$$

WORKED EXAMPLE: SLIDING DOWN A PLANE

Question: A block of mass $m = 3$ kg starts at rest at a height of h = 43 cm on a plane that has an angle of inclination of $\theta = 35^\circ$ with respect to the horizontal. The block slides down the plane, and, upon reaching the bottom, then slides along a horizontal surface. The coef_cient of kinetic friction of the block on both surfaces is $\mu = 0.25$. How far does the block slide along the horizontal surface before coming to rest?

Answer: The normal reaction of the plane to the block's weight is $R = mg \cos\theta$.

Hence, the frictional force acting on the block when it is sliding down the plane is $f = \mu R = 0.25 \times 3 \times 9.81 \times \cos 35^\circ = 6.03$ N.

The change in gravitational potential energy of the block as it slides down the plane is $\Delta U = -\, mgh = -\,3 \times 9.81 \times 0.43 = -12.65$ J. The work W done on the block by the frictional force during this process is,

$$W = -fx,$$

where $x = h/\sin\theta$ is the distance the block slides. The minus sign indicates that f acts in the opposite direction to the displacement of the block.

Hence,

$$W = -\frac{6.03 \times 0.43}{\sin 35°} = -4.52J.$$

Now, by energy conservation, the kinetic energy K of the block at the bottom of the plane equals the decrease in the block's potential energy plus the amount of work done on the block:

$$K = -\Delta U + W = 12.65 - 4.52 = 8.13 \text{ J}.$$

The frictional force acting on the block when it slides over the horizontal surface is,

$$f' = \mu\, mg = 0.25 \times 3 \times 9.81 = 7.36 \text{ N}.$$

The work done on the block as it slides a distance y over this surface is,

$$W' = -f' y.$$

By energy conservation, the block comes to rest when the action of the frictional force has drained all of the kinetic energy from the block: i.e., when $W' = -K$. It follows that,

$$y = \frac{K}{f'} = \frac{8.13}{7.36} = 1.10m.$$

WORKED EXAMPLE: DRIVING UP AN INCLINE

Question: A car of weight 3000 N possesses an engine whose maximum power output is 160 kW. The maximum speed of this car on a level road is 35 m/s. Assuming that the resistive force (due to a combination of friction and air resistance) remains constant, what is the car's maximum speed on an incline of 1 in 20 (i.e., if θ is the angle of the incline with respect to the horizontal, then $\sin\theta = 1/20$).

Answer: When the car is traveling on a level road at its maximum speed, v, then all of the power output, P, of its engine is used to overcome the power dissipated by the resistive force, f. Hence,

$$P = fv$$

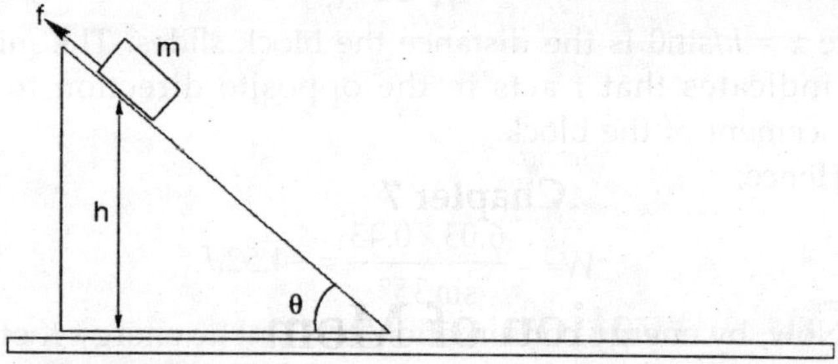

where the left-hand side is the power output of the engine, and the right-hand side is the power dissipated by the resistive force (i.e., minus the rate at which this force does work on the car). It follows that,

$$f = \frac{P}{v} = \frac{160 \times 10^3}{35} = 4.57 \times 10^3 N.$$

When the car, whose weight is W, is traveling up an incline, whose angle with respect to the horizontal is θ, it is subject to the additional force $f' = W \sin\theta$, which acts to impede its motion. Of course, this force is just the component of the car's weight acting down the incline. Thus, the new power balance equation is written,

$$P = fv' + W \sin\theta \, v',$$

where v' is the maximum velocity of the car up the incline. Here, the left-hand side represents the power output of the car, whereas the right-hand side represents the sum of the power dissipated by the resistive force and the power expended to overcome the component of the car's weight acting down the incline.

It follows that,

$$v' = \frac{P}{f + W \sin\theta} = \frac{160 \times 10^3}{1.57 \times 10^3 + 3000/20} = 33.90 m/s.$$

Chapter 7

Conservation of Momentum

The behaviour of dynamical systems which consist of single point masses (i.e., objects whose spatial extent is either negligible or plays no role in their motion) or arrangements of point masses which are constrained to move together because they are connected via inextensible cables.

Let us now broaden our approach somewhat in order to take into account systems of point masses which exert forces on one another, but are not necessarily constrained to move together. The classic example of such a multi-component point mass system is one in which two (or more) freely moving masses collide with one another. The physical concept which plays the central role in the dynamics of multi-component point mass systems is the conservation of momentum.

TWO-COMPONENT SYSTEMS

The simplest imaginable multi-component dynamical system consists of two point mass objects which are both constrained to move along the same straight-line. Let x_1 be the displacement of the first object, whose mass is m_1.

Likewise, let x_2 be the displacement of the second object, whose mass is m_2. Suppose that the first object exerts a force f_{21} on the second object, whereas the second object exerts a force f_{12} on the first. From Newton's third law of motion, we have,

$$f_{12} = -f_{21}.$$

Suppose, finally, that the first object is subject to an external force (i.e., a force which originates outside the system) F_1, whilst the second object is subject to an external force F_2.

Applying Newton's second law of motion to each object in turn, we obtain,

$$m_1 \ddot{x}_1 = f_{12} + F_1,$$
$$m_2 \ddot{x}_2 = f_{21} + F_2.$$

Here, is a convenient shorthand for d/dt. Likewise, means d^2/dt^2.

At this point, it is helpful to introduce the concept of the centre of mass. The centre of mass is an imaginary point whose displacement x_{cm} is defined to be the mass weighted average of the displacements of the two objects which constitute the system.

In other words,

$$x_{cm} = \frac{m_1 x_1 + m_2 x_2}{m_1 + m_2}$$

Thus, if the two masses are equal then the centre of mass lies half way between them, if the second mass is three times larger than the first then the centre of mass lies three-quarters of the way along the line linking the first and second masses, respectively, if the second mass is much larger than the first then the centre of mass is almost coincident with the second mass, and so on.

Summing Eqs. and, and then making use of Eqs. and, we obtain,

$$m_1 \ddot{x}_1 + m_2 \ddot{x}_2 = (m_1 + m_2)\, \ddot{x}_{cm} = F_1 + F_2.$$

Note that the internal forces, f_{12} and f_{21}, have canceled out. The physical significance of this equation becomes clearer if we write it in the following form,

$$M \ddot{x}_{cm} = F,$$

where $M = m_1 + m_2$ is the total mass of the system, and $F = F_1 + F_2$ is the net external force acting on the system. Thus, the motion of the centre of mass is equivalent to that which would occur if all the mass contained in the system were collected at the centre of mass, and this conglomerate mass were then acted upon by the net external force.

In general, this suggests that the motion of the centre of mass is simpler than the motions of the component masses, m_1 and m_2.

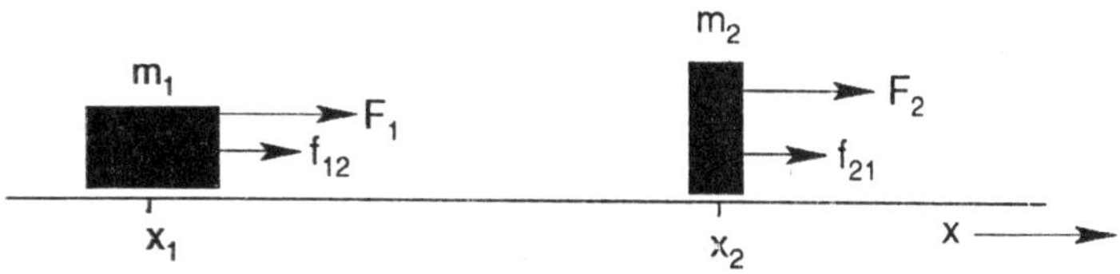

Fig. Dimensional Dynamical System Consisting of two point mass Objects

This is particularly the case if the internal forces, f_{12} and f_{21}, are complicated in nature.

Suppose that there are no external forces acting on the system (i.e., $F_1 = F_2 = 0$), or, equivalently, suppose that the sum of all the external forces is zero (i.e., $F = F_1 + F_2 = 0$). In this case, according to Eq., the motion of the centre of mass is governed by Newton's first law of motion. i.e., it consists of uniform motion in a straight-line. Hence, in the absence of a net external force, the motion of the centre of mass is almost certainly far simpler than that of the component masses.

Now, the velocity of the centre of mass is written,

$$v_{cm} = \dot{x}_{cm} = \frac{m_1\dot{x}_1 + m_2\dot{x}_2}{m_1 + m_2}.$$

We have seen that in the absence of external forces v_{cm} is a constant of the motion (i.e., the centre of mass does not accelerate). It follows that, in this case,

$$m_1\dot{x}_1 + \dot{x}_2\dot{x}_2$$

is also a constant of the motion. Momentum is defined as the product of mass and velocity. Hence, the momentum of the first mass is written $p_1 = m_1\dot{x}_1$, whereas the momentum of the second mass takes the form $p_2 = m_2\dot{x}_1$. It follows that the above expression corresponds to the total momentum of the system.

$$P = p_1 + p_2.$$

Thus, the total momentum is a conserved quantity—provided there is no net external force acting on the system. This is true irrespective of the nature of the internal forces. More generally, Eq. can be written

$$\frac{dP}{dt} = F.$$

In other words, the time derivative of the total momentum is equal to the net external force acting on the system—this is just Newton's second law of motion applied to the system as a whole.

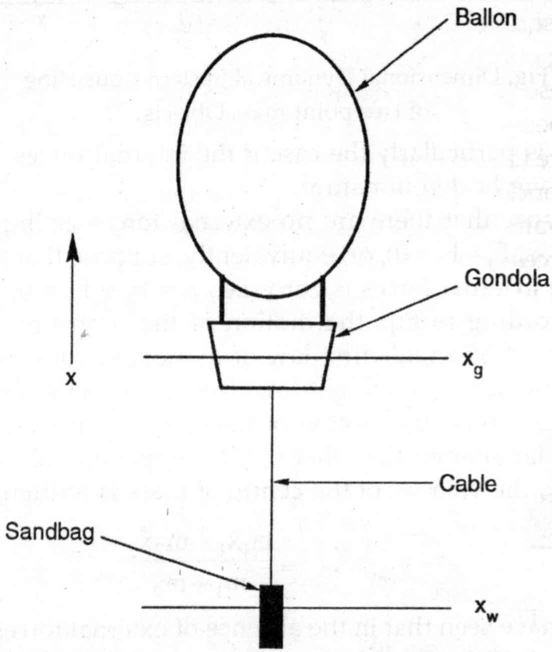

Fig: An Example Two-component System

Let us now try to apply some of the concepts discussed above. Consider the simple two-component system shown. A gondola of mass m_g hangs from a hot-air balloon whose mass is negligible compared to that of the gondola. A sandbag of mass m_w is suspended from the gondola by means of a light inextensible cable.

The system is in equilibrium. Suppose, for the sake of consistency with our other examples, that the x-axis runs vertically upwards. Let x_g be the height of the gondola, and x_w the height of the sandbag. Suppose that the upper end of the cable is attached to a winch inside the gondola, and that this winch is used to slowly shorten the cable, so that the sandbag is lifted upwards a distance Δx_w. The question is this.

does the height of the gondola also change as the cable is reeled in. If so, by how much?

Let us identify all of the forces acting on the system. The internal forces are the upward force exerted by the gondola on the sandbag, and the downward force exerted by the sandbag on the gondola.

These forces are transmitted via the cable, and are equal and opposite (by Newton's third law of motion). The external forces are the net downward force due to the combined weight of the gondola and the sandbag, and the upward force due to the buoyancy of the balloon. Since the system is in equilibrium, these forces are equal and opposite (it is assumed that the cable is reeled in sufficiently slowly that the,

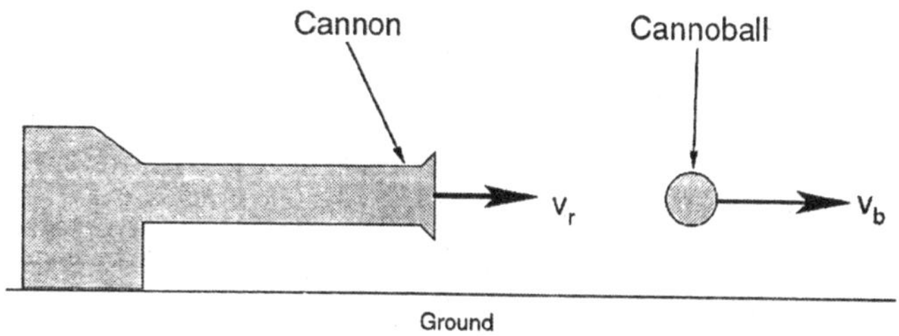

Fig: Another Example Two-component System

equilibrium is not upset). Hence, there is zero net external force acting on the system. It follows, from the previous discussion, that the centre of mass of the system is subject to Newton's first law. In particular, since the centre of mass is clearly stationary before the winch is turned on, it must remain stationary both during and after the period in which the winch is operated. Hence, the height of the centre of mass,

$$x_{cm} = \frac{m_g x_g + m_w x_w}{m_g + m_w},$$

is a conserved quantity.

Suppose that the operation of the winch causes the height of the sandbag to change by Δx_w, and that of the gondola to simultaneously change by Δx_g. If x_{cm} is a conserved quantity, then we must have,

$$0 = m_g \, \Delta x_g + m_w \, \Delta x_w,$$

or

$$\Delta x_g = \frac{m_w}{m_g} \Delta x_w .$$

Thus, if the winch is used to raise the sandbag a distance Δx_w then the gondola is simultaneously pulled downwards a distance $(m_w = m_g) \, \Delta x_w$.

It is clear that we could use a suspended sandbag as a mechanism for adjusting a hot-air balloon's altitude. the balloon descends as the sandbag is raised, and ascends as it is lowered.

Suppose that a cannon of mass M propels a cannonball of mass m horizontally with velocity v_b. What is the recoil velocity v_r of the cannon? Let us first identify all of the forces acting on the system.

The internal forces are the force exerted by the cannon on the cannonball, as the cannon is Δred, and the equal and opposite force exerted by the cannonball on the cannon. These forces are extremely large, but only last for a short instance in time. in physics, we call these impulsive forces.

There are no external forces acting in the horizontal direction (which is the only direction that we are considering in this example). It follows that the total (horizontal) momentum P of the system is a conserved quantity. Prior to the Firing of the cannon, the total momentum is zero (since momentum is mass times velocity, and nothing is initially moving). After the cannon is Δred, the total momentum of the system takes the form,

$$P = m v_b + M v_r .$$

Since P is a conserved quantity, we can set P = 0.

Hence,

$$v_r = -\frac{m}{M} v_b .$$

Thus, the recoil velocity of the cannon is in the opposite direction to the velocity of the cannonball (hence, the minus sign in the above equation), and is of magnitude $(m = M) \, v_b$.

Of course, if the cannon is far more massive that the cannonball (i.e., M >> m), which is usually the case, then the recoil velocity of the cannon is far smaller in magnitude than the velocity of the cannonball.

Note, however, that the momentum of the cannon is equal in magnitude to that of the cannonball.

It follows that it takes the same effort (i.e., force applied for a certain period of time) to slow down and stop the cannon as it does to slow down and stop the cannonball.

MULTI-COMPONENT SYSTEMS

Consider a system of N mutually interacting point mass objects which move in 3-dimensions. Let the ith object, whose mass is m_i, be located at vector displacement r_i. Suppose that this object exerts a force f_{ji} on the jth object.

By Newton's third law of motion, the force f_{ij} exerted by the jth object on the ith is given by,

$$f_{ij} = -f_{ji}.$$

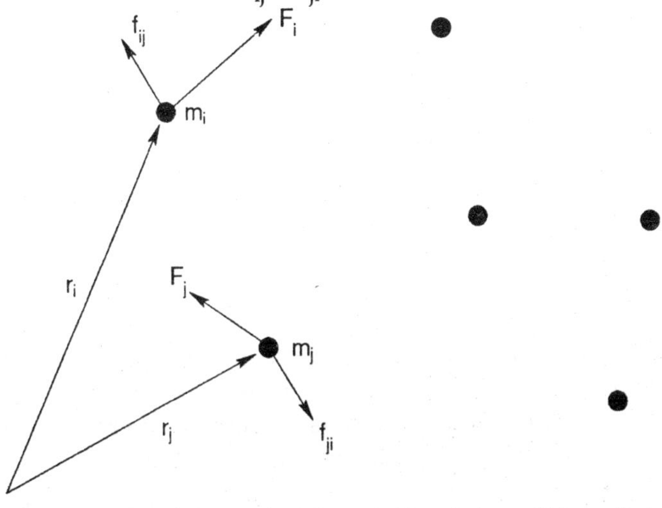

Fig: Dimensional Dynamical System Consisting of Many Point Mass Objects.

Finally, suppose that the ith object is subject to an external force F_i. Newton's second law of motion applied to the ith object yields,

$$m_i \ddot{r}_i = \sum_{j=1,N}^{j \neq i} f_{ij} + F_i .$$

Note that the summation on the right-hand side of the above equation excludes the case j = i, since the ith object cannot exert a force on itself. Let us now take the above equation and sum it over all objects.

We obtain,

$$\sum_{i=1,N} m_i \ddot{r}_i \sum_{i=1,N}^{j \neq i} f_{ij} + \sum_{i=1,N} F_i .$$

Consider the sum over all internal forces. i.e., the first term on the right-hand side.

Each element of this sum—f_{ij}, say—can be paired with another element— f_{ji}, in this case—which is equal and opposite. In other words, the elements of the sum all cancel out in pairs. Thus, the net value of the sum is zero.

It follows that the above equation can be written,

$$M \ddot{r}_{cm} = F,$$

where,

$$M = \sum_{i=1}^{N} m_i$$

is the total mass, and,

$$F = \sum_{i=1}^{N} F_i$$

is the net external force.

The quantity r_{cm} is the vector displacement of the centre of mass. As before, the centre of mass is an imaginary point whose coordinates are the mass weighted averages of the coordinates of the objects which constitute the system.

Thus,

$$r_{cm} = \frac{\sum_{i=1}^{N} m_i r_i}{\sum_{i=1}^{N} m_i}$$

The motion of the centre of mass is equivalent to that

which would be obtained if all the mass contained in the system were collected at the centre of mass, and this conglomerate mass were then acted upon by the net external force. As before, the motion of the centre of mass is likely to be far simpler than the motions of the component masses.

Suppose that there is zero net external force acting on the system, so that F = 0. In this case, Eq. implies that the centre of mass moves with uniform velocity in a straight-line. In other words, the velocity of the centre of mass,

$$\dot{r}_{cm} = \frac{\sum_{i=1}^{N} m_i \dot{r}_i}{\sum_{i=1}^{N} m_i},$$

is a constant of the motion. Now, the momentum of the ith object takes the form $p_i = m_i \dot{r}_i$. Hence, the total momentum of the system is written

$$P = \sum_{i=1}^{N} m_i \dot{r}_i$$

A comparison of Eqs. and suggests that P is also a constant of the motion when zero net external force acts on the system. Finally, Eq. can be rewritten

$$\frac{dP}{dt} = F.$$

In other words, the time derivative of the total momentum is equal to the net external force acting on the system.

It is clear, from the above discussion, that most of the important results obtained in the previous section, for the case of a two-component system moving in 1-dimension, also apply to a multi-component system moving in 3-dimensions.

As an illustration of the points raised in the above discussion, let us con-sider the unfortunate history of the planet Krypton. As you probably all know, Krypton—Superman's home planet—eventually exploded. Note, however, that before, during, and after this explosion the net external force acting on Krypton, or the fragments of

Krypton—namely, the gravitational attraction tc Krypton's sun—remained the same.

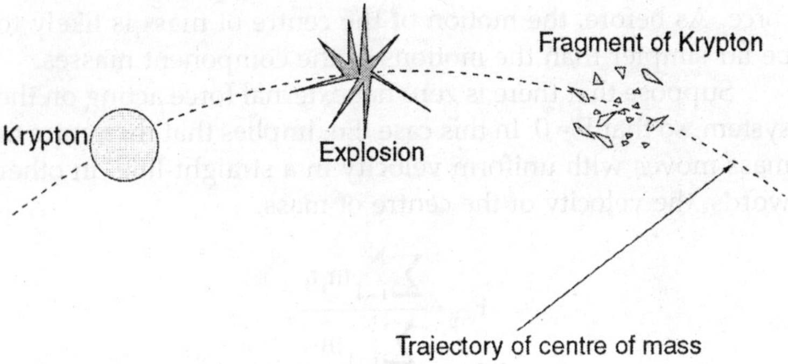

Fig: The Unfortunate History of the Planet Krypton.

In other words, the forces responsible for the explo-sion can be thought of as large, transitory, internal forces. We conclude that the motion of the centre of mass of Krypton, or the fragments of Krypton, was unaffected by the explosion. This follows, since the motion of the centre of mass is independent of internal forces. Before the explosion, the planet Krypton presumably executed a standard elliptical orbit around Krypton's sun.

We conclude that, after the explosion, the fragments of Krypton (or, to be more exact, the centre of mass of these fragments) continued to execute exactly the same orbit.

ROCKET SCIENCE

A rocket engine is the only type of propulsion device that operates effectively in outer space. A rocket works by ejecting a propellant at high velocity from its rear end. The rocket exerts a backward force on the propellant, in order to eject it, and, by Newton's third law, the propellant exerts an equal and opposite force on the rocket, which propels it forward.

Let us attempt to find the equation of motion of a rocket. Let M be the fixed mass of the rocket engine and the payload, and m(t) the total mass of the propellant contained in the rocket's fuel tanks at time t. Suppose that the rocket engine ejects the propellant at some fixed velocity u relative to the rocket.

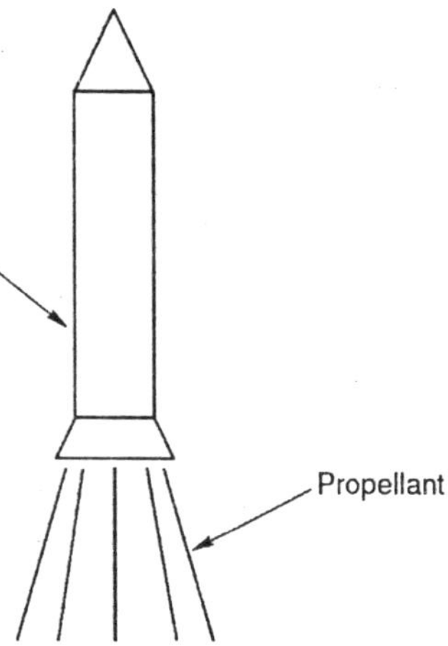

Fig: A Rocket.

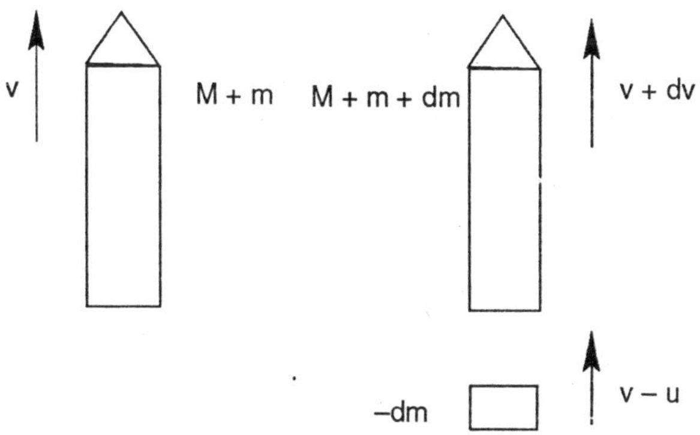

Fig: Derivation of the Rocket Equation.

Let us examine the rocket at two closely spaced instances in time. Suppose that at time t the rocket and propellant, whose total mass is M + m, are traveling with instantaneous velocity v. Suppose, further, that between times t and t + dt the rocket ejects a quantity of propellant of mass –dm (n.b., dm is understood to be negative, so this represents a positive mass) which travels with velocity v – u (i.e., velocity -u in the instantaneous rest frame of the rocket). As a result of the fuel ejection, the velocity of the rocket at time t + dt is boosted to v

+ dv, and its total mass becomes M + m + dm. Now, there is zero external force acting on the system, since the rocket is assumed to be in outer space. It follows that the total momentum of the system is a constant of the motion. Hence, we can equate the momenta evaluated at times t and t + dt.

$$(M + m)v = (M + m + dm)(v + dv) + (-dm)(v - u).$$

Neglecting second order quantities (i.e., dm dv), the above expression yields,

$$0 = (M + m)dv + udm.$$

Rearranging, we obtain,

$$\frac{dv}{u} = -\frac{dm}{M+m}.$$

Let us integrate the above equation between an initial time at which the rocket is fully fueled—i.e., $m = m_p$, where m_p is the maximum mass of propellant that the rocket can carry—but stationary, and a final time at which the mass of the fuel is m and the velocity of the rocket is v.

Hence,

$$\int_0^v \frac{dv}{u} = -\int_{m_p}^{m} \frac{dm}{M+m}.$$

It follows that,

$$\left[\frac{v}{u}\right]_{v=0}^{v=v} = -[\ln(M+m)]_{m=m_p}^{m=m},$$

which yields,

$$\mu = u \ln\left(\frac{M+m_p}{M+m}\right).$$

The final velocity of the rocket (i.e., the velocity attained by the time the rocket has exhausted its fuel, so that m = 0) is

$$v_f = u \ln\left(1+\frac{m_p}{M}\right)$$

Note that, unless the initial mass of the fuel exceeds the fixed mass of the rocket by many orders of magnitude (which is highly unlikely), the final velocity v_f of the rocket is similar to the velocity u with which fuel is ejected from the rear of the rocket in its instantaneous rest frame. This follows because ln x ~ O(1), unless x becomes extremely large.

Let us now consider the factors which might influence the design of a rocket for use in interplanetary or interstellar travel. Since the distances involved in such travel are vast, it is important that the rocket's final velocity be made as large as possible, otherwise the journey is going to take an unacceptably long time. However, as we have just seen, the factor which essentially determines the final velocity v_f of a rocket is the speed of ejection u of the propellant relative to the rocket.

Broadly speaking, v_f can never significantly exceed u. It follows that a rocket suitable for interplanetary or interstellar travel should have as high an ejection speed as practically possible.

Now, ordinary chemical rockets (the kind which powered the Apollo moon programme) can develop enormous thrusts, but are limited to ejection velocities below about 5000 m/s. Such rockets are ideal for lifting payloads out of the Earth's gravitational field, but their relatively low ejection velocities render them unsuitable for long distance space travel.

A new type of rocket engine, called an ion thruster, is currently under development. ion thrusters operate by accelerating ions electrostatically to great velocities, and then ejecting them.

Although ion thrusters only generate very small thrusts, compared to chemical rockets, their much larger ejection velocities (up to 100 times those of chemical rockets) makes them far more suitable for interplanetary or interstellar space travel. The first spacecraft to employ an ion thruster was the Deep Space 1 probe, which was launched from Cape Canaveral on October 24, 1998. this probe successfully encountered the asteroid 9969 Braille in July, 1999.

COLLISIONS IN 1-DIMENSION

Consider two objects of mass m_1 and m_2, respectively, which are free to move in 1-dimension. Suppose that these two objects collide. Suppose, further, that both objects are subject to zero net force when they are not in contact with one another. This situation is illustrated.

Both before and after the collision, the two objects move with constant velocity. Let v_{i1} and v_{i2} be the velocities of the first and second objects, respectively, before the collision.

Likewise, let v_{f1} and v_{f2} be the velocities of the first and second objects, respectively, after the collision. During the collision itself, the first object exerts a large transitory force f_{21} on the second, whereas the second object exerts an equal and opposite force $f_{12} = -f_{21}$ on the first. In fact, we can model the collision as equal and opposite impulses given to the two objects at the instant in time when they come together.

We are clearly considering a system in which there is zero net external force (the forces associated with the collision are internal in nature). Hence, the total momentum of the system is a conserved quantity. Equating the total momenta before and after the collision, we obtain,

$$m_1\, vi_1 + m_2\, vi_2 = m_1\, vf_1 + m_2\, vf_2.$$

This equation is valid for any 1-dimensional collision, irrespective its nature. Note that, assuming we know the masses of the colliding objects, the above equation only fully describes the collision if we are given the initial velocities of both objects, and the final velocity of at least one of the objects. (Alternatively, we could be given both final velocities and only one of the initial velocities.)

There are many different types of collision. An elastic collision is one in which the total kinetic energy of the two colliding objects is the same before and after the collision. Thus, for an elastic collision we can write,

$$\frac{1}{2}m_1 v_{i1}^2 + \frac{1}{2}m_2 v_{i2}^2 = \frac{1}{2}m_1 v_{f1}^2 + \frac{1}{2}m_2 v_{f2}^2,$$

in addition to Eq.. Hence, in this case, the collision is fully speci∆ed once we are given the two initial velocities of the colliding objects. (Alternatively, we could be given the two final velocities.)

The majority of collisions occurring in real life are not elastic in nature. Some fraction of the initial kinetic energy of the colliding objects is usually converted into some other form of energy—generally heat energy, or energy associated with the mechanical deformation of the objects—during the collision.

Such collisions are termed inelastic. For instance, a large fraction of the initial kinetic energy of a typical automobile accident is converted into mechanical energy of deformation of the two vehicles. Inelastic collisions also occur during squash/racquetball/handball games. in each case, the ball becomes warm to the touch after a long game, because some fraction of the ball's kinetic energy of collision with the walls of the court has been converted into heat energy. Equation remains valid for inelastic collisions—however, Eq. is invalid. Thus, generally speaking, an inelastic collision is only fully characterized when we are given the initial velocities of both objects, and the final velocity of at least one of the objects.

There is, however, a special case of an inelastic collision—called a totally inelastic collision—which is fully characterized once we are given the initial velocities of the colliding objects. In a totally inelastic collision, the two objects stick together after the collision, so that $v_{f1} = v_{f2}$.

Let us, now, consider elastic collisions in more detail. Suppose that we transform to a frame of reference which co-moves with the centre of mass of the system. The motion of a multi-component system often looks particularly simple when viewed in such a frame. Since the system is subject to zero net external force, the velocity of the centre of mass is invariant, and is given by,

$$v_{cm} = \frac{m_1 v_{i1} + m_2 v_{i2}}{m_1 + m_2} = \frac{m_1 v_{f1} + m_2 v_{f2}}{m_1 + m_2}$$

An object which possesses a velocity v in our original frame of reference—henceforth, termed the laboratory frame—possesses a velocity $v' = v - v_{cm}$ mass frame. It is easily demonstrated that,

$$v'_{i1} = -\frac{m_2}{m_1 + m_2}(v_{i2} - v_{i1}),$$

$$v'_{i2} = +\frac{m_1}{m_1 + m_2}(v_{i2} - v_{i1})$$

$$v'_{f1} = -\frac{m_2}{m_1 + m_2}(v_{f2} - v_{f1})$$

$$v'_{f2} = +\frac{m_1}{m_1 + m_2}(v_{f2} - v_{f1}).$$

The above equations yield,

$$-p'_{i1} = p'_{i2} = \mu\,(v_{i2} - v_{i1}),$$

$$-p'_{f1} = p'_{f2} = \mu(v_{f2} - v_{f1}),$$

where $\mu = m_1\ m_2 = (m_1 + m_2)$ is the so-called reduced mass, and $-p'_{i1} = m_1 v'_{i1}$ is the initial momentum of the first object in the centre of mass frame, etc. In other words, when viewed in the centre of mass frame, the two objects approach one another with equal and opposite momenta before the collision, and diverge from one another with equal and opposite momenta after the collision. Thus, the centre of mass momentum conservation equation,

$$p'_{i1} + p'_{i2} = p'_{f1} + p'_{f2},$$

is trivially satisfied, because both the left- and right-hand sides are zero. Incidentally, this result is valid for both elastic and inelastic collisions.

The centre of mass kinetic energy conservation equation takes the form,

$$\frac{p'^2_{i1}}{2m_1} + \frac{p'^2_{i2}}{2m_2} = \frac{p'^2_{f1}}{2m_1} + \frac{p'^2_{f2}}{2m_2}.$$

Note, incidentally, that if energy and momentum are conserved in the laboratory frame then they must also be conserved in the centre of mass frame. A comparison of Eqs.,, and yields

$$(v_{i2} - v_{i1}) = -(v_{f2} - v_{f1}).$$

In other words, the relative velocities of the colliding

objects are equal and opposite before and after the collision. This is true in all frames of reference, since relative velocities are frame invariant. Note, however, that this result only applies to fully elastic collisions. Equations and can be combined to give the following pair of equations which fully specify the final velocities (in the laboratory frame) of two objects which collide elastically, given their initial velocities.

$$v_{f1} = \frac{(m_1 - m_2)}{m_1 + m_2} v_{i1} + \frac{2m_2}{m_1 + m_2} v_{i2},$$

$$v_{f2} = \frac{2m_1}{m_1 + m_2} v_{i1} - \frac{(m_1 - m_2)}{m_1 + m_2} v_{i2}.$$

Let us, now, consider some special cases. Suppose that two equal mass objects collide elastically. If $m_1 = m_2$ then Eqs. and yield

$$v_{f1} = v_{i2},$$
$$v_{f2} = v_{i1}.$$

In other words, the two objects simply exchange velocities when they collide. For instance, if the second object is stationary and the first object strikes it head-on with velocity v then the first object is brought to a halt whereas the second object moves off with velocity v. It is possible to reproduce this effect in pool by striking the cue ball with great force in such a manner that it slides, rather that rolls, over the table— in this case, when the cue ball strikes another ball head-on it comes to a complete halt, and the other ball is propelled forward very rapidly. Incidentally, it is necessary to prevent the cue ball from rolling, because rolling motion is not taken into account in our analysis, and actually changes the answer.

Suppose that the second object is much more massive than the first (i.e., $m_2 \gg m_1$) and is initially at rest (i.e., $v_{i2} = 0$). In this case, Eqs. and yield

$$v_{f1} = -v_{i1},$$
$$v_{f2} = 0.$$

In other words, the velocity of the light object is effectively reversed during the collision, whereas the massive object remains approximately at rest. Indeed, this is the sort of behaviour we expect when an object collides elastically with an immovable obstacle. e.g., when an elastic ball bounces off a brick wall. Suppose, finally, that the second object is much lighter than the first (i.e., $m_2 \ll m_1$) and is initially at rest (i.e., $v_{i2} = 0$). In this case, Eqs. and yield

$$v_{f1} = v_{i1},$$
$$v_{f2} = 2\, v_{i1}.$$

In other words, the motion of the massive object is essentially unaffected by the collision, whereas the light object ends up going twice as fast as the massive one. Let us, now, consider totally inelastic collisions in more detail.

In a totally inelastic collision the two objects stick together after colliding, so they end up moving with the same final velocity $v_f = v_{f1} = v_{f2}$. In this case, Eq. reduces to

$$v_f = \frac{m_1 v_{i1} + m_2 v_{i2}}{m_1 + m_2} = v_{cm}.$$

In other words, the common final velocity of the two objects is equal to the centre of mass velocity of the system. This is hardly a surprising result. We have already seen that in the centre of mass frame the two objects must diverge with equal and opposite momenta after the collision. However, in a totally inelastic collision these two momenta must also be equal (since the two objects stick together).

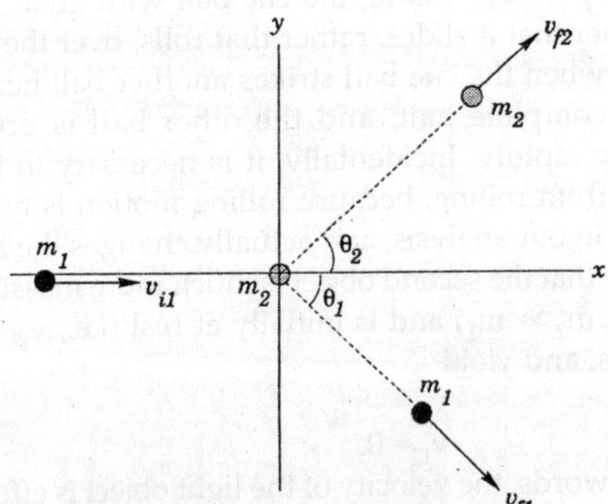

Fig: A Collision in 2-dimensions.

The only way in which this is possible is if the two objects remain stationary in the centre of mass frame after the collision. Hence, the two objects move with the centre of mass velocity in the laboratory frame. Suppose that the second object is initially at rest (i.e., $v_{i2} = 0$). In this special case, the common final velocity of the two objects is,

$$v_f = \frac{m_1}{m_1 + m_2} v_{i1} \,.$$

Note that the first object is slowed down by the collision. The fractional loss in kinetic energy of the system due to the collision is given by,

$$f = \frac{K_i - K_f}{K_i} = \frac{m_1 v_{i1}^2 - (m_1 + m_2) v_f^2}{m_1 v_{i1}^2} = \frac{m_2}{m_1 + m_2} \,.$$

The loss in kinetic energy is small if the (initially) stationary object is much lighter than the moving object (i.e., if $m_2 \ll m_1$), and almost 100% if the moving object is much lighter than the stationary one (i.e., if $m_2 > m_1$). Of course, the lost kinetic energy of the system is converted into some other form of energy. e.g., heat energy.

COLLISIONS IN 2-DIMENSIONS

Suppose that an object of mass m_1, moving with initial speed v_{i1}, strikes a second object, of mass m_2, which is initially at rest. Suppose, further, that the collision is not head-on, so that after the collision the first object moves off at an angle θ_1 to its initial direction of motion, whereas the second object moves off at an angle θ_2 to this direction. Let the final speeds of the two objects be v_{f1} and v_{f2}, respectively.

We are again considering a system in which there is zero net external force (the forces associated with the collision are internal in nature). It follows that the total momentum of the system is a conserved quantity. However, unlike before, we must now treat the total momentum as a vector quantity, since we are no longer dealing with 1-dimensional motion. Note that if the collision takes place wholly within the x-y plane, as indicated in Figure then it is sufΔcient to equate the x- and y-components of the total momentum before and after the collision. Consider the x-component of the system's total momentum. Before the collision, the total x-momentum is simply $m_1 v_{i1}$, since the second object is initially stationary, and the first object is initially moving along the x-axis with speed v_{i1}. After the collision, the x-momentum of the first object is m1 $v_{f1} \cos \theta_1$. i.e., m1 times the x-component of the first object's final velocity. Likewise, the final xmomentum of the second object is $m_2 v_{f2} \cos \theta_2$. Hence, momentum conservation in the x-direction yields,

$$m_1 v_{i1} = m_1 v_{f1} \cos \theta_1 + m_2 v_{f2} \cos \theta_2.$$

Consider the y-component of the system's total momentum. Before the collision, the total y-momentum is zero, since there is initially no motion along the y-axis. After the collision, the y-momentum of the first object is $-m_1 v_{f1} \sin \theta_1$. i.e., m_1 times the y-component of the first object's final velocity. Likewise, the final y-momentum of the second object is $m_2 v_{f2} \sin \theta_2$.

Hence, momentum conservation in the y-direction yields,

$$m_1 v_{f1} \sin \theta_1 = m_2 v_{f2} \sin \theta_2.c$$

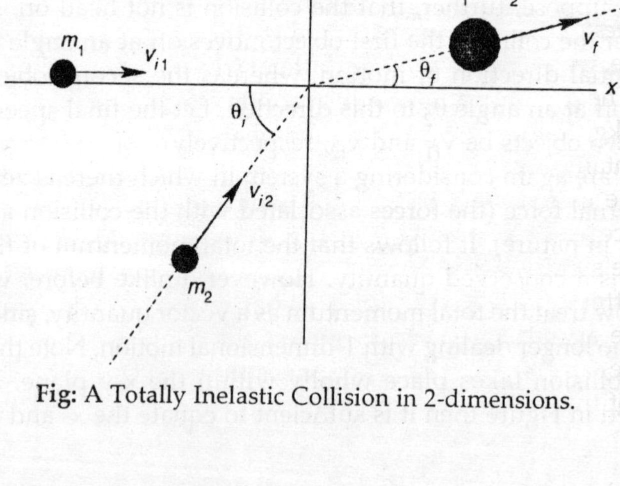

Fig: A Totally Inelastic Collision in 2-dimensions.

For the special case of an elastic collision, we can equate the total kinetic energies of the two objects before and after the collision.

Hence, we obtain,

$$\frac{1}{2}m_1 v_{i1}^2 = \frac{1}{2}m_1 v_{f1}^2 + \frac{1}{2}m_2 v_{f2}^2.$$

Given the initial conditions (i.e., m_1, m_2, and v_{i1}), we have a system of three equations [i.e., Eqs.,, and] and four unknowns (i.e., θ_1, θ_2, v_{f1}, and v_{f2}). Clearly, we cannot uniquely solve such a system without being given additional information. e.g., the direction of motion or speed of one of the objects after the collision.

A 2-dimensional totally inelastic collision. In this case, the first object, mass m_1, initially moves along the x-axis with speed v_{i1}. On the other hand, the second object, mass m_2, initially moves at an angle θ_i to the x-axis with speed v_{i2}. After the collision, the two objects stick together and move off at an angle θ_f to the x-axis with speed v_f. Momentum conservation along the x-axis yields,

$$m_1 v_{i1} + m_2 v_{i2} \cos \theta_i = (m_1 + m_2) v_f \cos \theta_f.$$

Likewise, momentum conservation along the y-axis gives,

$$m_2 v_{i2} \sin \theta_i = (m_1 + m_2) v_f \sin \theta_f.$$

Given the initial conditions (i.e., m_1, m_2, v_{i1}, v_{i2}, and θ_i), we have a system of two equations [i.e., Eqs. and] and two unknowns (i.e., v_f and θ_f). Clearly, we should be able to find a unique solution for such a system.

Question: A cannon is bolted to the Θoor of a railway carriage, which is free to move without friction along a straight track. The combined mass of the cannon and the carriage is M = 1200 kg. The cannon fires a cannonball, of mass m = 1.2 kg, horizontally with velocity v = 115 m/s.

The cannonball travels the length of the carriage, a distance L = 85 m, and then becomes embedded in the carriage's end wall. What is the recoil speed of the carriage right after the cannon is fired? What is the velocity of the carriage after the cannonball strikes the far wall? What net distance, and in what direction, does the carriage move as a result of the firing of the cannon?

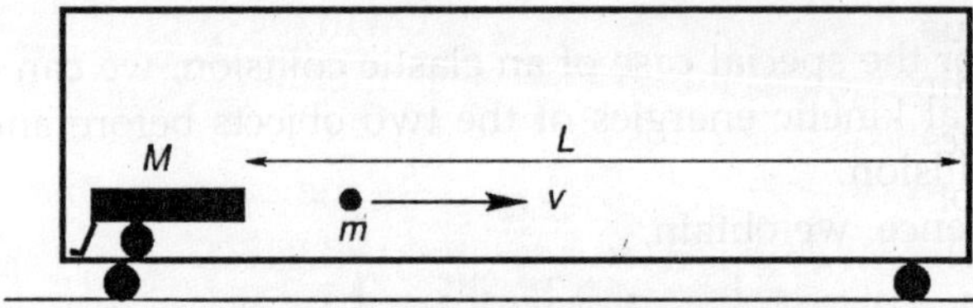

Answer: Conservation of momentum implies that the net horizontal momentum of the system is the same before and after the cannon is fired. The momentum before the cannon is fired is zero, since nothing is initially moving. Hence, we can also set the momentum after the cannon is fired to zero, giving

$$0 = Mu + mv,$$

where u is the recoil velocity of the carriage. It follows that,

$$u = -\frac{m}{M}v = -\frac{1.2 \times 115}{1200} = -0.115 \text{m/s}$$

The minus sign indicates that the recoil velocity of the carriage is in the opposite direction to the direction of motion of the cannonball. Hence, the recoil speed of the carriage is $|u| = 0.115$ m/s.

Suppose that, after the cannonball strikes the far wall of the carriage, both the cannonball and the carriage move with common velocity w. Conservation of momentum implies that the net horizontal momentum of the system is the same before and after the collision.

Hence, we can write,

$$Mu + mv = (M + m)w.$$

However, we have already seen that $M u + mv = 0$. It follows that $w = 0$. in other words, the carriage is brought to a complete halt when the cannonball strikes its far wall.

In the frame of reference of the carriage, the cannonball moves with velocity v – u after the cannon is fired. Hence, the time of Θight of the cannonball is,

$$t = \frac{L}{v - u} = \frac{85}{115 + 0.115} - 0.738\text{s}.$$

The distance moved by the carriage in this time interval is,

$$d = ut = -0.115 \times 0.738 = -0.0849 \text{ m}.$$

Thus, the carriage moves 8.49 cm in the opposite direction to the direction of motion of the cannonball.

Question: A softball of mass m = 0.35 kg is pitched at a speed of u = 12 m/s. The batter hits the ball directly back to the pitcher at a speed of v = 21 m/s. The bat acts on the ball for t = 0.01 s. What impulse is imparted by the bat to the ball? What average force is exerted by the bat on the ball?

Answer: The initial momentum of the softball is -m u, whereas its final momentum is mv. Here, the final direction of motion of the softball is taken to be positive. Thus, the net change in momentum of the softball due to its collision with the bat is,

$$\Delta p = mv - (-)mu = 0.35 \times (21 + 12) = 11.55 \text{ Ns}.$$

By definition, the net momentum change is equal to the impulse imparted by the bat, so

$$I = \Delta p = 11.55 \text{ N s}.$$

The average force exerted by the bat on the ball is simply the net impulse divided by the time interval over which the ball is in contact with the bat. Hence,

Worked Example. Skater and Medicine Ball

Question: A skater of mass M = 120 kg is skating across a pond with uniform velocity v = 8 m=s. One of the skater's friends, who is standing at the edge of the pond, throws a medicine ball of mass m = 20 kg with velocity u = 3 m=s to the skater, who catches it.

The direction of motion of the ball is perpendicular to the initial direction of motion of the skater. What is the final speed of the skater? What is the final direction of motion of the skater relative to his/her initial direction of motion? Assume that the skater moves without friction.

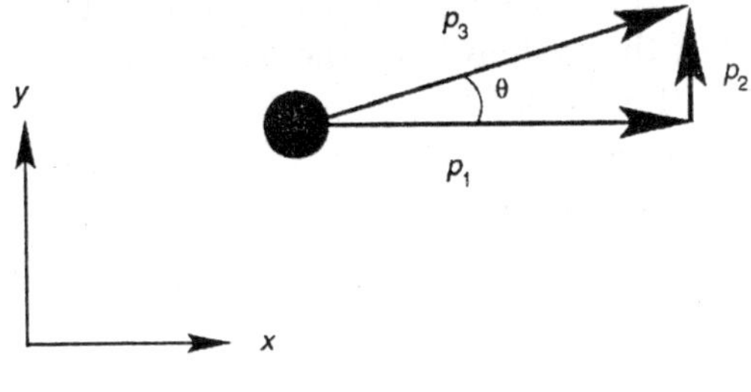

Answer: Suppose that the skater is initially moving along the x-axis, whereas the initial direction of motion of the medicine ball is along the y-axis. The skater's initial momentum is,

$$p_1 = (M\ v, 0) = (120 \times 8, 0) = (960, 0)\ \text{N s}.$$

Likewise, the initial momentum of the medicine ball is,

$$p_2 = (0, m\ u) = (0, 20 \times 3) = (0, 60)\ \text{N s}.$$

After the skater catches the ball, the combined momentum of the skater and the ball is,

$$p_3 = p_1 + p_2 = (960, 60)\ \text{Ns}.$$

This follows from momentum conservation. The final speed of the skater (and the ball) is,

$$v' \frac{|p_3|}{M+m} = \frac{\sqrt{960^2 + 60^2}}{120+20} = 6.87\ \text{m/s}.$$

The final direction of motion of the skater is parametreized by the angle θ, where,

$$\theta = \tan^{-1}\left(\frac{|p_2|}{|p_1|}\right) = \tan^{-1}\left(\frac{60}{960}\right) = 3.58°$$

Worked Example Bullet and Block

Question: A bullet of mass m = 12 g strikes a stationary wooden block of mass M = 5.2 kg standing on a frictionless surface. The block, with the bullet embedded in it, acquires a velocity of v = 1.7 m=s.

What was the velocity of the bullet before it struck the block? What fraction of the bullet's initial kinetic energy is lost (i.e., dissipated) due to the collision with the block?

Answer: Let u be the initial velocity of the bullet. Momentum conservation requires the total horizontal momentum of the system to be the same before and after the bullet strikes the block.

The initial momentum of the system is simply mu, since the block is initially at rest. The final momentum is (M + m) v, since both the block and the bullet end up moving with velocity v.

Hence,

$$mu = (M + m)v,$$

giving,

$$u = \frac{M+m}{m}v = \frac{(0.012+5.2)\times 1.7}{0.012} = 738.4 m/s.$$

The initial kinetic energy of the bullet is,

$$K_i = \frac{1}{2}mu^2 = 0.5\times 0.012\times 738.4^2 = 3.2714 kJ.$$

The final kinetic energy of the system is,

$$K_i = \frac{1}{2}(M+m)v^2 = 0.5\times(0.012+5.2)\times 1.7^2 = 7.53 J$$

Hence, the fraction of the initial kinetic energy which is dissipated is,

$$f = \frac{K_i - K_f}{K_i} = \frac{3.2714\times 10^3 - 7.53}{3.2714\times 10^3} = 0.9977.$$

Worked Example Elastic Collision

Question: An object of mass m_1 = 2 kg, moving with velocity v_{i1} = 12 m=s, collides head-on with a stationary object whose mass is m_2 = 6 kg. Given that the collision is elastic, what are the final velocities of the two objects. Neglect friction.

Answer: Momentum conservation yields,

$$m_1 v_{i1} = m_1 v_{f1} + m_2 v_{f2},$$

where v_{f1} and v_{f2} are the final velocities of the first and second objects, respectively. Since the collision is elastic, the total kinetic energy must be the same before and after the collision.

Hence,

$$\frac{1}{2}m_1 v_{i1}^2 + \frac{1}{2}m_1 v_{f1}^2 + \frac{1}{2}m_2 v_{f1}^2.$$

Let $x = v_{f1}/v_{i1}$ and $y = v_{f2}/v_{i1}$. Noting that $m_2/m_1 = 3$, the above two equations reduce to

$$1 = x + 3y,$$

and

$$1 = x^2 + 3y^2.$$

Eliminating x between the previous two expressions, we obtain,

$$1 = (1 - 3y)^2 + 3y^2,$$

or

$$6y(2y - 1) = 0,$$

which has the non-trivial solution y = 1/2. The corresponding solution for x is x = (1 – 3y) = –1/2.

It follows that the final velocity of the first object is,

$$v_{f1} = x\, v_{i1} = -0.5 \times 12 = -6 \text{ m/s}.$$

The minus sign indicates that this object reverses direction as a result of the collision. Likewise, the final velocity of the second object is,

$$v_{f2} = y\, v_{i1} = 0.5 \times 12 = 6 \text{ m/s}.$$

Worked Example 2-dimensional Collision

Question. Two objects slide over a frictionless horizontal surface. The first object, mass $m_1 = 5$ kg, is propelled with speed $v_{i1} = 4.5$ m=s toward the second object, mass $m_2 = 2.5$ kg, which is initially at rest. After the collision, both objects have velocities which are directed θ = 30° on either side of the original line of motion of the first object. What are the final speeds of the two objects? Is the collision elastic or inelastic?

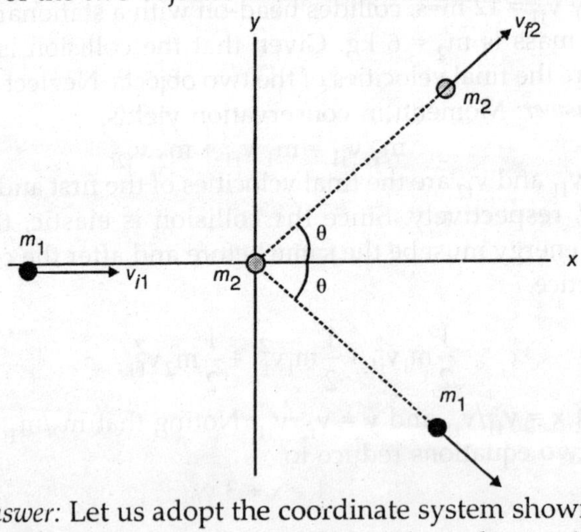

Answer: Let us adopt the coordinate system shown in the diagram. Conservation of momentum along the x-axis yields

$$m_1 \, v_{i1} = m_1 \, v_{f1} \cos \theta + m_2 \, v_{f2} \cos \theta.$$

Likewise, conservation of momentum along the y-axis yields,

$$m_1 \, v_{f1} \sin \theta = m_2 \, v_{f2} \sin \theta.$$

The above pair of equations can be combined to give

$$v_{f1} = \frac{v_{i1}}{2\cos\theta} = \frac{4.5}{2 \times \cos 30°} = 2.5981 \text{ m/s},$$

and

$$v_{f2} = \frac{m_1}{m_2} v_{f1} = \frac{5 \times 5.981}{2.5} = 5.1962 \text{ m/s}.$$

The initial kinetic energy of the system is,

$$K_i = \frac{1}{2} m_1 v_{i1}^2 = 0.5 \times 5 \times 4 \times 4.5^2 = 50.63\text{J}.$$

The final kinetic energy of the system is,

$$K_f = \frac{1}{2} m_1 v_{f1}^2 + \frac{1}{2} m_2 v_{f2}^2$$

$$= 0.5 \times 5 \text{ Q } 2.5981^2 + 0.5 \times 2.5 \times 5.1962^2 = 50.63 \text{ J}.$$

Since $K_i = K_f$, the collision is elastic.

Chapter 8

Kinetics

CONCEPT

Newton's formulation of mechanics, which involved the new concepts of mass and force, was subjected to intense, but sterile, criticism by Ernst Mach (1838-1916) and others, which did not change its application to the slightest degree, and shed no light on its fundamentals.

The ideas of Albert Einstein (1879-1955), on the other hand, modified the fundamentals essentially, especially in the theory of gravity and the kinematics of motion, making the theory able to explain the smallest discrepancies with the earlier predictions.

This *theory of relativity* has also been abundantly proved by observation. Except for Newton and Einstein, all the vast amount of work in mechanics has been the elaboration and application of the theory, which has proved in every respect correct.

Particularly notable is the work of Joseph Louis Lagrange (1736-1813) in The Beautiful Theory, where forces are replaced by work and energy, and the laws of mechanics can be expressed as problems in maxima and minima. This branch of mechanics is usually called *analytical mechanics*, and mechanics with forces is called *vector mechanics*, from the convenient use of that formalism.

Newton's theory of motion, in its original form, is still the basis of engineering mechanics, and a fundamental part of engineering training. Indeed, it is still used in celestial mechanics. The only aspect needing relativistic corrections so far has been the time effects in the Global Positioning System. In the solar system, velocities are very comfortably less than the speed of light, 3×10^8 m/s, although distances are great. Light requires about 8 minutes to pass from the sun to the earth, and gravity travels at the same speed.

However, the sun does not care a great deal about the force exerted on it by the earth, while the earth moves in a static gravitational field, so the effects of the finite speed of gravity are not evident.

These effects have, however, recently been measured and verified, so it is known that gravity is not instantaneous. We will not need to consider any relativistic effects in the sequel, leaving these tiny effects to another article.

NEWTON'S LAWS

Newton's laws of motion are:

- A body under no forces moves uniformly in a straight line;
- The acceleration of a body is equal to the force acting on it divided by its mass;
- The mutual forces acting between two bodies are equal and opposite.

Newton stated the laws in this form so that these *axioms* would use only common notions and not involve any results following from them, in the spirit of Euclidean geometry.

The first law gives an operational method for determining an *inertial system*, a way of describing the position of a mass point in space as a function of time. In the absence of gravitating masses (such as the earth) it is found that a Euclidean space is such a system, in which a uniform time can be approximated by periodic events (such as the oscillations in lasers).

Strictly (and uselessly) speaking, the definition is circular, but in practice it is clearly possible to establish inertial systems at least approximately, and the results obtained agree perfectly with observation.

The second law defines force as the cause of a time rate of change of velocity, by $dv/dt = f/m$. Again, the definition is logically faulty, but can be realised with great accuracy in practice. Whatever causes a change in velocity in an inertial system is a force.

Masses can be compared dynamically, or more easily by the gravitational forces on them. A consistent array of masses and forces can be assembled, which always gives the observed motion.

When the velocity is known, the position can be found from the kinematical relation dr/dt = v. Relations not involving force we call *kinematical,* those involving force, *dynamical.*

The third law guarantees that the total linear momentum and total angular momentum will be conserved; that is, that they will remain constant during the motion.

These quantities are defined later, and their conservation is the motive for the third law, which cannot be stated in a general form in other than an extremely tedious and unilluminating way.

Of course, not all forces obey it (magnetic forces between currents are an example), but nevertheless momenta are conserved. The conservation of momentum is an important fundamental of mechanics; the third law is a way of introducing it in terms of forces. In celestial mechanics, all the forces are central (act along the line joining two interacting masses) and so the simple statement is sufficient.

The path traced by the tip of the velocity vector v as time elapses is called the *hodograph,* and the path traced by the tip of the position vector r is called the *trajectory*. An orbit is a special kind of trajectory, a closed (or almost closed) curve. These curves are useful in describing, visualizing and analyzing the motion.

The *momentum* is the product p = mv. The second law can be written dp/dt = f, so that force is the rate of change of momentum. If two particles m and m′ interact, then f' = -f from the third law, so that (d/dt)(p + p') = f + f' = 0, or dP = 0, where P = p + p' is the total momentum. The two particles, taken as a whole, constitute an isolated system, on which no net force acts, so the rate of change of momentum is zero, and thus the momentum is constant.

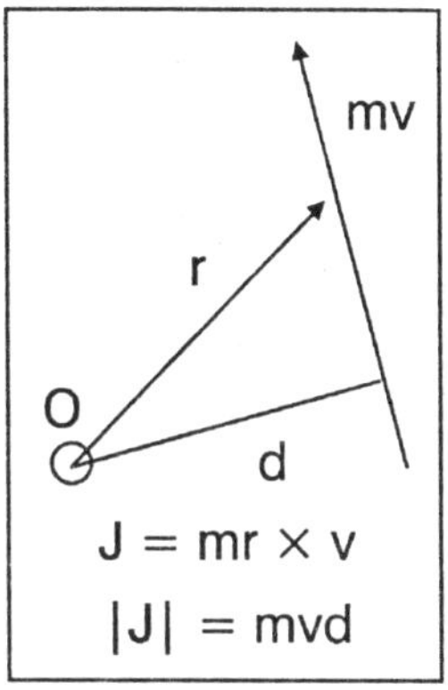

Fig. Angular Momentum

The *angular momentum* with respect to a point O is the vector J = r × p, where r is the radius vector from O to any point on the line of action of p = mv. In the figure, J points out of the page. If we use r x p instead of p in the equations of the preceding paragraph, we can prove that the total angular momentum is conserved. The conservation of linear and angular momentum gives us six scalar constants of the motion, useful in analyzing and understanding the motion. The existence of these constants is no trivial matter, but a beautiful and fundamental result of the theory.

The second law can be formally integrated to give p = ∫ fdt = I, where I is a vector called the *impulse,* the integral of the force with respect to time. By taking moments about point O, we obtain a similar expression for the angular impulse. Of course, the limits on the integrals must be handled properly.

If the second law is scalar-multiplied by v, we find mv·(dv/dt) = f·v, or (d/dt)($mv^2/2$) = f·v = -dW/dt. This introduces two new scalar quantities, the *kinetic energy* T = $mv^2/2$, and the *work* -W = ∫ f·dr. The minus sign, is simply means that work is being done *on* the particle, while W is conventionally the work done *by* the particle. If W is a function of position, then in vector notation f = – grad W. In one dimension, f = – dW/dx, which may be more familiar. What we have shown is that (d/dt)(T + W) = 0, or T + W = constant, a new conservation law. If W is a *conservative* quantity (that is, its value is independent of path and depends only on position), it is called the *potential energy* V. Then, the *total energy* E = T + V is conserved in the motion. Energy is a widely used and misused quantity that could be discussed in great detail, but only its bare definition is required here, and the reader is assumed to be familiar with it.

The introduction of energy permits the use of *generalized coordinates*, and the derivation of equations of motion by the Lagrange procedure, which uses the Lagrangian function L = T– V. This facilitates the solutions of very many problems, since we are liberated from carrying around a basket of vector components.

For vector mechanics, it is convenient to have expressions for the velocity and acceleration components in polar coordinates.

These are derived in the References, but will be summarized here. v = (dr/dt)r′ + (rdθ/dt)θ′. r′ and θ′ are unit vectors pointing radially outward, and tangentially in the anticlockwise sense, with respect to the position vector r, as shown in the diagram. The rectangular unit vectors i and j are also shown. Note that dr′/dθ = θ′ and dθ′/dθ = –r′, while the derivatives with respect to r are zero. dv/dt = $[d^2r/dt^2 - r(d\theta/dt)^2]r' + (1/r)(d/dt)(r^2 d\theta/dt)\theta'$.

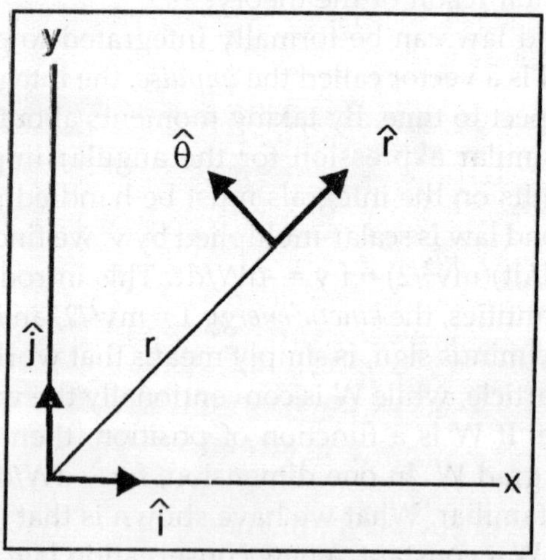

Fig: Unit Vectors

The reader should carefully note the distinction between, for example, d^2r/dt^2 and $d^2\mathbf{r}/dt^2$. The first is the second derivative of a scalar quantity, a simple thing.

The second is a vector, and since the direction of r can change, the complete expression is somewhat elaborate, as we have just seen. It is essential to distinguish carefully between scalars and vectors, which are arrays of scalars.

These relations are very useful. For example, suppose that the force is central, so that the tangential component of the acceleration is zero. This means that $(d/dt)(r^2d\theta/dt) = 0$, or $r^2d\theta/dt$ = constant = h. Now, half of r times r $d\theta$ is the area dA swept out by the radius vector in dt, so h is 2dA/dt. Also, $mrd\theta/dt$ is the component of the momentum perpendicular to the radius vector, so $mr^2d\theta/dt = mh$ is the angular momentum.

We then have that the angular momentum L = mh is a constant. Also, this means that the motion takes place in a plane, and L is perpendicular to this plane. We have found out a lot here with very little effort. The constancy of areal velocity is one of the most useful properties of orbital motion, not just a curiosity. It relates the rate of change of the angle θ to the radius r.

KEPLERIAN ORBITS

Kepler's three laws of planetary motion, deduced by prolonged and tedious consideration of the observed position of Mars, are:

- The planets move in ellipses with the sun at one focus;
- The areas swept out by the radius vector in equal time intervals are equal; and
- The cubes of the mean distances (half the major axis of the orbit) are proportional to the squares of the periodic times. These laws are sufficient to determine the position of a planet at any later time if its position is known at one time, and the dimensions and orientation of the orbit are known.

For a deeper understanding, and the power to attack an arbitrary problem in orbital motion, such as the movement of earth satellites, we should consider the dynamics, on the basis of Newton's theory.

The force acting between two spherical, radially symmetric bodies of masses M and m a distance r apart is, by Newtonian gravitation, GMm/r^2 in magnitude, and is directed along the line joining the centres of the two bodies. G is the Newtonian gravitational constant, 6.67259×10^{-11} $m^3/kg\text{-}s$. Since the force is central, the angular momentum is conserved, and the bodies revolve about one another in a plane with the centre of mass (the barycentre) fixed. The motion can be analysed as the rotation of a *reduced mass* Mm/(M + m) about the centre of gravity. The potential energy V is -GMm/r, and is negative because the particles attract. For the total energy to be zero, $T = mv^2/2 = GMm/r$, or $v = \Phi(2GM/r)$.

This is the *escape velocity* from the particle of mass M at the distance r. At the surface of the earth, the escape velocity is 11,200 m/s or 25,200 mph. This is also the velocity of a meteor that comes in from infinity. The escape velocity from the *sun* at the distance of the earth is 42,000 m/s or about 95,000 mph. It is not easy to get away from the earth; it is even harder to escape from the sun.

If we consider a circular orbit of radius a, as shown in the figure at the left, then $GMm/a^2 = [Mm/(M + m)]a(d\theta/dt)^2$, since $v = a(d\theta/dt)$ and the centripetal acceleration is v^2/a. This is the equation of motion for the relative motion, with r the distance between M and m. The centre of gravity of M and m moves with constant velocity, not affected by the relative motion. Thus, $G(M + m) = a^3(d\theta/dt)^2 = 4\pi^2(a^3/P^2)$, where P is the period of the motion.

This is a derivation of Kepler's third law for a special case, but it is a general result when a is the mean distance of any orbit. The quantity G(M + m) is usually called μ, so that $r^3/P^2 = \mu/4\pi^2$. If r is in astronomical units (radius of the earth's orbit, about 150×10^6 km), and the period P is in years, then $G(M + m) = 4\pi^2 = \mu$. The quantity GM is called the Gaussian gravitational constant k^2 if time is in days instead of years, so we get approximately $k^2 = 4\pi^2/(365.25)^2$, and k = 0.017202. A more precise value is 0.01720209895. Untangling G from GM means knowing the mass of the sun accurately, which is impossible, so celestial mechanics uses k instead of G. The value should be corrected for the finite mass of the earth, but since the ratio of the sun's to the earth's mass is 332,946 the correction is small.

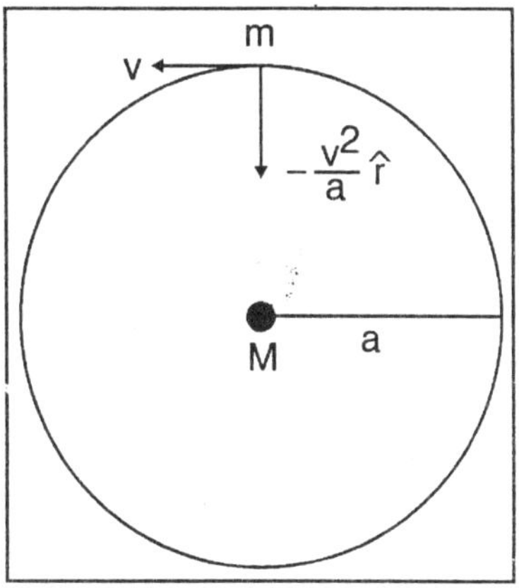

Fig: Circular Orbit

The polar equation of an ellipse, with the origin in a focus, is $r = p/(1 + e \cos \theta)$. The *eccentricity* of the ellipse is e, and p is a length called the semi-latus-rectum. In connection with orbits, the angle θ is called the *true anomaly*. In an elliptical orbit, 0 d" e < 1. When e = 0, we have a circular orbit r = p. The same formula gives a parabola for e = 1, and a hyperbola for e > 1. In an ellipse, the maximum and minimum radii are $p/(1 - e)$ and $p/(1 + e)$, respectively, and the corresponding points on the orbit are called *aphelion* and *perihelion*, respectively. The distance from perihelion to aphelion is the major axis of the ellipse, twice the mean distance a.

Therefore, $2a = p/(1 - e) + p/(1 + e) = 2p/(1 - e^2)$, or $p = a(1 - e^2)$. The perihelion distance is $a(1 - e)$, the aphelion distance $a(1 + e)$. The distance from the centre to a focus is $a - a(1 - e) = ea = c$. This defines the eccentricity as $e = c/a$. Since the ellipse is the locus of points the sum of whose distances to the two foci is a constant, 2a, the length of the radius from a focus to the end of the minor axis is a. Since this is the hypotenuse of a right triangle, $a^2 = c^2 + b^2$, where b is the semiminor axis. Hence

$b = a\,\Phi(1 - e^2)$. The reader can show that the average value of r over one revolution is a. This is about all we need to know about the geometry of the ellipse for present purposes.

If we know the orbit is an ellipse, and the areal velocity is constant, we should be able to prove that the force is central and varies as the inverse square of the distance. On the other hand, if we know that the force is central and inverse-square, then we should be able to prove that the orbit is an ellipse and the areal velocity is constant. Either of these things can be done fairly easily.

In the first case, differentiate the expression for r with respect to time, obtaining $dr/dt = (e/p) \sin \theta\, 2A$, where A is the areal velocity (this was called dA/dt above; now A represents the derivative). Then differentiate again to find $d^2r/dt^2 = (4A^2e/pr^2) \cos \theta$. The tangential acceleration is zero from the constancy of the areal velocity. The radial acceleration is $a_r = (4A^2e/pr^2) \cos \theta - 4A^2/r^3 = -4A^2/pr^2$, which shows that the force is inverse square, and we have already shown that it is central. In fact, $GM = 4A^2/p$.

In the second case, the equations of motion in the plane of the orbit are $d^2r/dt^2 - r(d\theta/dt)^2 = -GM/r^2$ and $rd^2\theta/dt + 2(dr/dt)(d\theta/dt) = 0$.

Now, $(d/dt)(r^2d\theta/dt) = 2r(dr/dt)(d\theta/dt) + r^2d^2\theta/dt^2 = 0$, so the areal velocity $A = (1/2)r^2(d\theta/dt) = \text{constant}$.

Using this result, the radial equation becomes $d^2r/d\theta^2 - (2/r)(dr/d\theta)^2 - r = -r^2GM/4A^2$. Making the substitution $u = 1/r$, this equation can be thrown into the form $d^2u/d\theta^2 + u = GM/4A^2$. This second-order linear equation with constant coefficients is easy to solve. We find $u = 1/r = GM/4A^2 + C \cos \theta$. This is the equation of a conic section, as we have seen above. If $p = 4A^2/GM$ and $e = pC$, we have $r = p/(1 + e \cos \theta)$, which is what we wanted to prove.

Now we can find the components of the velocity. The transverse component is $rd\theta/dt = 2A/r = 2A(1 + e \cos \theta)/p$, where A is the areal velocity. The radial component is $dr/dt = 2A(e/p)\sin \theta$.

The square of the velocity is the sum of the squares, or $v^2 = 4A^2[(1 + e^2 + 2e \cos \theta]/p^2] = 4A^2[(1 - e^2)/p^2 + 2/pr] = -4A^2/pa +$

$4A^2/pr$, from which the kinetic energy per unit mass (i.e., setting m = 1 for simplicity) is: $v^2/2 = -GM/2a + GM/r$, since $4A^2/p = GM$.

Rearranging, $v^2/2 - GM/r = -GM/2a$, which is to say, T + V = E, with $T = v^2/2$, $V = -GM/r$ and $E = -GM/2a$. The total energy is negative (i.e., the orbiting particle is "trapped") and a function of the mean distance a only. The eccentricity is determined by the angular momentum h through $e^2 = 1 - h^2/aGM$. When h = "aGM, the orbit is circular.

The quantity a, called the "mean distance," is not the average value of the radius vector r. It is the average of the perihelion and aphelion distances, however. The time average value of 1/r turns out to be 1/a, which is its real definition.

This means that the time average potential energy is -GM/a, and the time average kinetic energy is -GM/2a + GM/a = +GM/2a. For a circular orbit, the kinetic and potential energies are constant. The time average value of r is $a(1 + e^2/2)$, and the aver2age of r over the true anomaly is b, the minor axis of the orbit. The average of 1/r over the true anomaly is 1/p.

We obtained the relation between the mean distance a and the orbital period P for the special case of a circular orbit. A more general proof is as follows. The areal velocity dA/dt = h/2 = constant, so integrating from t = 0 to t = P we find A = hP/2 = πab. We also know that $p = a(1 - e^2) = h^2/GM$. Therefore, $P^2 = 4\pi^2a^2b^2/h^2 = [4\pi^2/GM]a^3$, since $b^2/p = a$. This is the desired general result. If we know the period and mean distance of any orbit, we can calculate the gravitational constant GM, which in the general case is G(M + m).

COORDINATES AND ORBITS

The angular coordinates that are used to describe directions in space are shown in the figure at the left. The equator and ecliptic planes are shown. The dihedral angle between these two planes is $\varepsilon = 23° 26' 21''.412$, or thereabouts. The equatorial coordinates are the right ascension α and the declination δ, while the analogous ecliptic coordinates are the longitude λ and the latitude β. The relations between the coordinates are best found by considering the rectangular components of the vector OP, considered as of unit length, and then performing the rotation about the x-axis that takes one system into the other. We will not require these relations at the present time.

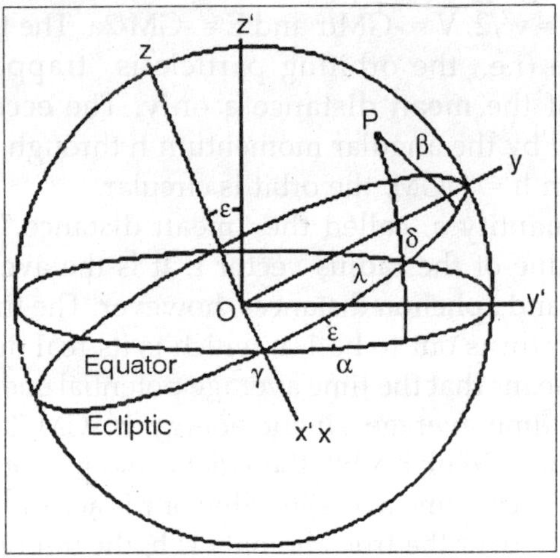

Fig: Equatorial and Ecliptic Coordinates

The position of a body in the solar system can be specified in terms of a gigantic rectangular coordinate system centreed on the sun. The x-axis is always taken as pointing to the vernal equinox, a direction specified by the line of intersection of the orbital plane of the earth (the ecliptic plane) with a plane parallel to the equator of the earth, the equatorial plane.

Both contain the centre of the sun. This is made interesting by the precession of the normal to the equatorial plane around the pole of the ecliptic. In the diagram, this is the movement of z′ (the pole of the equator) around z (the pole of the ecliptic). Fortunately, this is not too rapid, taking about 26,500 years for one revolution.

The equatorial plane has to be considered because our earth-based measurements use it as a reference, when directions are given in terms of right ascension and declination. If we know a vector whose components are referred to this plane, we can find the right ascension and declination of its direction. When the earth is located in its orbital plane on the line of intersection of the ecliptic and equatorial planes, the sun is seen in front of the vernal equinox about March 21, climbing from below the equator to above it. At this time, when we look at the sun (careful!) we are looking in the direction of the x-axis of the heliocentric coordinate system.

The y-axis of the ecliptic coordinates is in the plane of the ecliptic, at right angles to the x-axis, and in the direction in which the planets move, in which a screw would advance in the direction of the north pole of the earth.

The ecliptic z-axis is then perpendicular to the x- and y-axes, and makes a right-handed coordinate system with them. The equatorial rectangular coordinates are defined analogously. If we represent these coordinates by primes, × and x′ are the same, while x′,y′ is rotated with respect to x,y by the dihedral angle between the planes. The rotation from the z′ axis to the z axis is a right-handed rotation about the positive x,x′ axis.

Longitudes are measured clockwise (eastwards) on the ecliptic plane from the vernal equinox, from 0° to 360°. The *longitude of perihelion*, ψ, is the angle measured clockwise from the vernal equinox to the radius passing through perihelion. (A different symbol, a kind of ω, is used conventionally, but it is not available here in HTML.) The longitude of perihelion of the earth's orbit is currently about 103°. This locates the major axis of the orbit and the direction from which the true anomaly is measured.

Other planets move on orbital planes inclined to the ecliptic by the *inclination*, i. They intersect the ecliptic on a line called the *line of nodes*. The *longitude of the ascending node*, Ω, is the angle measured clockwise from the vernal equinox to the line of nodes, on the end where the orbit is climbing above the ecliptic plane, calleu the *ascending node*. The node 180° away is, not surprisingly, the *descending node*. Note in the figure that the traditional symbol for the ascending node is used, which we represent in text by Ω, which looks similar. The symbol for the descending node is inverted, with the two little loops on the top. The longitude of perihelion in this case is the longitude of the ascending node, *plus* the angle ω in the orbital plane from the node to the perihelion. This may be a little odd, but it does locate the perihelion. That is, $\psi = \Omega + \omega$. These quantities are illustrated in the diagram.

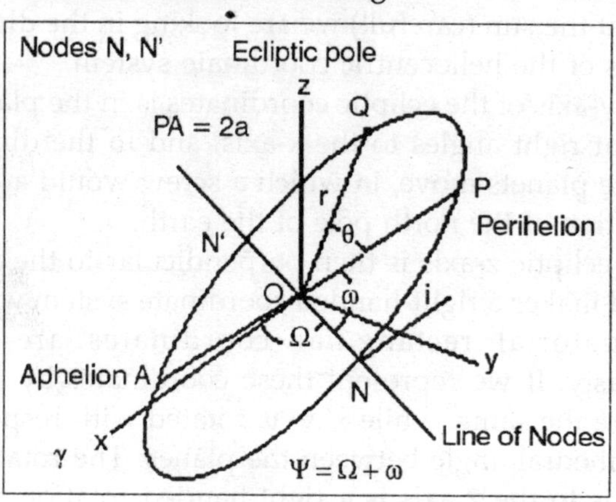

Fig: Orbital Elements

Now that the orbital plane and the direction of the perihelion have been specified by the three orbital elements i, Ω and ψ, we proceed to specify the shape of the orbit by its mean distance a and its eccentricity e. The position of the body in its orbit at a specified time is given by a sixth element, the mean longitude at the epoch.

The "epoch" is just the reference time assumed, usually by its Julian date. The mean distance a determines the orbital period P through Kepler's third law. Therefore, only one of a or P need be specified. The *mean daily motion,* 360° divided by the orbital period, is often given in place of P. Orbital elements for planets, asteroids and comets can be found online at the Jet Propulsion Laboratory website whose link is given in the References.

The constancy of the areal velocity is used to locate the body in its orbit at a specified time. This is called Kepler's Problem, which Kepler also solved. The solution is illustrated in the figure. A reference circle of radius a is drawn around the orbital ellipse. Point Q is projected onto the auxilary circle by a perpendicular to the major axis to point Q'. The line CQ' then defines the angle E at the centre of the circle, called the *eccentric anomaly.* We can show that it is related to the radius vector r and the true anomaly θ by $r = a(1 - e \cos E)$ and $\tan(\theta/2) = \Phi[(1 + e)/(1 - e)] \tan E/2$. The shaded sector PFQ is the area swept out by the radius vector, and the area is proportional to the time. The area FPQ′ is proportional to the area FPQ, since it is merely magnified vertically in the ratio $a/b = 1/\Phi(1 - e^2)$. Indeed, the total area of the orbit is πab, while the total area of the circle is πa^2.

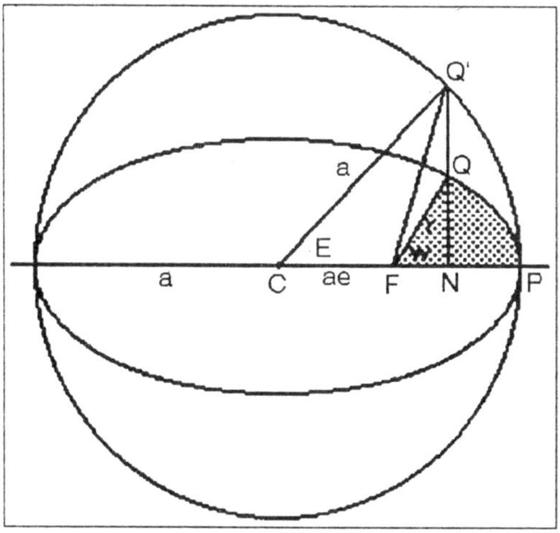

Fig: Kepler's Problem

Therefore, this larger area FPQ′ increases uniformly with time as well. Since the total area is πa^2, the area swept out is given by $(\pi a^2/P)t$, where t is the time since perihelion passage. This area can also be found in terms of E, since it is the area of the whole sector PCQ′ less the area of the triangle CQ′F.

The area of the sector is $\pi a^2(E/2\pi) = (a^2/2)E$. The area of the triangle is $(1/2)(a \sin E)(ae) = (a^2/2)e \sin E$. Therefore, we have $E - e \sin E = 2\pi t/P = M$, called Kepler's Equation. M is called the *mean anomaly* and increases proportionally to time after perihelion passage. This is really a beautiful result, allowing us to find the position in orbit in terms of a uniformly increasing quantity.

Kepler's Equation can be easily solved by iteration. If it is written $E = M + e \sin E$, we find the iteration equation $E_{n+1} = M + e \sin E_n$ (e is the eccentricity here, not the base of natural logarithms). Starting with $E_0 = M$, a few interations are enough to get good precision, especially when e is small.

The iterations are very easy, and can be performed on a pocket calculator. For e = 0.1, six iterations give 8 digits. The Newton-Raphson method can also be used. This method finds the roots of the equation f(x) = 0 by improving an initial guess x_0 by the formula $x_{n+1} = x_0 - f(x_n)/f'(x_n)$. A sketch will reveal the reasoning behind this formula. In this case, f(E) = M – E + e sin E = 0 is the function, and f′(E) = e cos E – 1. Fewer iterations are required, but each requires more work, than in the case of simple iteration. In practice, an expansion in powers of e is often used for small e, but the iteration methods are no more trouble, especially for a computer.

Although the determination of position in orbit may sound complicated, it is really quite straightforward, and after you have done it a few times, it will be easy. Let's find the position of the earth on 2003 April 2. The figures come from the *Astronomical Almanac for 2003*, p. E4. On Jan 1, JD 245 2640.5, the mean longitude of the earth was 100.2440°, and the longitude of perihelion was 103.0°. April 2 is 91 days later, so M = 100.2440 – 103.0 + 91 × 0.9856 = 86.93° = 1.5172 radians. A first approximation to E is M + e sin M, and e = 0.0167, so the correction is 0.01668 radians, and E = 1.5339 radians or 87.89°. There is no need to iterate again.

The mean distance can be taken as 1.00000, so r = 0.9994 and θ = 88.84°. The earth's longitude is 103.0 + 88.84 = 191.84°. On p. C8, the solar ephemeris gives longitude 11.8506° and r = 0.9994. The solar longitude is 180° less than the earth's, or 11.84° according to our calculation. In both cases, we are using the mean ecliptic of date, and came quite close using Keplerian orbital elements instead of the more elaborate calculations used to calculate the solar ephemeris. Precise orbits require the consideration of planetary perturbations, and this is a very difficult subject indeed.

The orbit of the moon is an example, since it is subject to perturbations by the earth's spheroidal shape and the sun that make its orbit vary in a very complicated way. In the case of the earth's orbit, we really calculate the position of the earth-moon barycentre. This point is 4671 km from the centre of the earth, about two-thirds of the way to the surface.

The mean distance of the moon is 3.844×10^5 km, and its sidereal period is 27.321661 days, from which G(M + m) can be calculated and used to relate mean distance and period for other earth satellites after correcting for the mass of the moon. The moon's radius is 1738 km, and its mass 7.3483×10^{22} kg. m/M = 0.0123, which is not quite negligible. The eccentricity of the moon's orbit is about 0.0549, and its inclination 5.145396°.

The plane of the orbit and the position of perigee change rather rapidly. The small eccentricity of the earth's orbit suggests that the earth-moon system did not suffer any external disturbance during its formation, and in particular any collision which created the moon. This was a one-time event, so speculations on what happened can never be proved, since no similar state is known, or is ever likely to become known.

Except for Mercury and Pluto, orbital inclinations are less than 4°, and eccentricities less than 0.0936, the value for Mars. Kepler was fortunate to have picked Mars for study because of its relatively large eccentricity, about twice those of Jupiter and Saturn. Venus's orbit is practically circular, with e = 0.0067.

Mercury not only has a large inclination, 7.005°, but an eccentric orbit, with e = 0.2056. Only Pluto has a greater inclination and eccentricity, but it is probably a special case. Observations of Mercury and Pluto were not available to Kepler, who had to work with Tycho's naked-eye observations.

The angle made by r with the x-axis in our problem was 191.84°, so the coordinates will be x = 0.9994 cos 191.84° = -0.97814, and y = 0.9994 x sin 191.84° = –0.20506, in astronomical units. In the more general case of a planet in an inclined orbital plane, the radius is first projected on the line of nodes [$a = r \cos(\omega + \theta)$] and on a perpendicular (dip) line [$b = r \sin(\omega + \theta)$]. Then $z = b \sin i$, $y = b \cos i \cos \Omega + a \sin \Omega$, $x = a \cos \Omega - b \cos i \sin \Omega$. It is just a matter of projecting r on the coordinate axes.

The necessary formulas are given on page E4 of the almanac. There is an expansion for the true anomaly directly in terms of the mean anomaly, without going through the eccentric anomaly.

To find the equatorial rectangular components, rotate about the x-axis through an angle of i. This gives $y' = y \cos i - z \sin i$ and $z' = y \sin i + z \cos i$. If x', y' and z' are the components of a vector in equatorial rectangular components, then the right ascension α is given by $\cos \alpha = x'/''(x'^2 + y'^2)$, and the declination δ by $\sin \delta = z'/''(x'^2 + y'^2 + z'^2)$. The proper quadrants have to be determined. In this way, the direction in which a planet is seen from the earth can be found.

COMETARY ORBITS

Comets can be divided into two classes, the *short-period* comets, like Halley's Comet, with a period of 76 years, and *long-period* comets, like Hale-Bopp (1995 O1), which do not return for thousands of years, if at all. A period of 200 years is the conventional dividing-line. Short-period comets have elliptical orbits like the planets, except that the eccentricity is larger, the inclination can take any value, and the comets can move in a *direct* or *retrograde* direction. Retrograde motion is usually expressed by an inclination greater than 90°.

Comets were the first real test for Newton's theory, which finally showed that they were normal members of the solar system, not mysterious atmospheric happenings, as had always been supposed. Positions of short-period comets are calculated in the same way as planetary positions, and the orbital elements are presented the same way.

Long-period comets come from the periphery of the solar system, where they wander in the hypothetical Oort Cloud of cometary debris, normally water and carbon dioxide ice and dust. Their total energy is about zero, so the eccentricity of their orbits when they make an excursion towards the sun is about 1.

That is, their orbits are parabolic. Short-period comets are those that have suffered an energy-losing collison with a planet (usually Jupiter) and have dropped into an elliptical orbit. There must also have been comets that have gained energy, entered hyperbolic orbits, and were ejected from the solar system. Few, if any, comets have eccentricities significantly greater than 1, which would indicate that they were encountered by the solar system in its path through space, and are not part of the family.

The parabola is a considerably simpler orbit than the ellipse. Its polar formula is $r = p/(1 + \cos\theta) = (p/2)\sec^2(\theta/2)$. The areal velocity $A = \Phi(GMp/4)$. Setting $x = \tan(\theta/2)$, direct integration gives Kepler's Equation as $x + x^3/3 = \Phi(GM/2)(2/p)^{3/2}(t - T)$. The one real root of this equation gives the position in orbit. There are many ways to solve a cubic equation: refer to mathematics handbooks algebra texts or mathematics programmes. The time to move from perihelion to the end of the latus rectum is $t_1 = (1/2)''(p^3/GM)$. $2t_1$ is a good measure of the time that the comet will spend near the sun.

The orbital elements include the inclination i, longitude of the ascending node Ω, and argument of perihelion ω. The perihelion distance p/2 is given, and the JD of perihelion passage. For Hale-Bopp, the perihelion distance was 0.91399384 and the time of perihelion passage was JD 245 0539.60742. The inclination was 89.42064850°, longitude of ascending node 282.47215310°, and argument of perihelion 130.59561740. From these elements, it was possible to find the position of the comet as seen from the earth.

Cometary orbital elements are available on the internet soon after the discovery of a comet. The Jet Propulsion Laboratory supplied the elements for Hale-Bopp shortly after it was discovered.

Halley's comet was the first periodic comet to be recognized. Edmund Halley (1656-1742), friend of Newton's and later Astronomer Royal, suspected that the comets of 1531 and 1607 were the same as the comet of 1682, since they had similar orbits. He predicted the return of the comet in 1758, which duly occurred. Halley had published Newton's *Principia* at his own expense in 1678, and made an extensive study of comets using the new theory. Halley's comet is the brightest of the short-period comets; most are quite dim, and can even have orbits of small eccentricity like asteroids.

To show how to use the orbital elements given by JPL,

let's look at them for Halley's comet. The time of perihelion passage, T_p is shown as 19860209.45895, which we interpret to mean 1986 February 9, at 0.45895 part of the day, or 11h 0m 53s UT. The "epoch" is the date to which the elements apply, which is given as 46480. This is a modified JD. The JD can be obtained by adding 24400000.5, or JD 2446480.5, which is 1986 February 18.

The size and shape of the orbit is specified by the perihelion distance q = 0.58710374, in AU, and e = 0.96727724. The eccentricity is nearly, but not quite, unity. Near perihelion, the orbit will be indistinguishable from a parabola, so the parabolic formulas can be used for predicting its position while it is near the sun and visible. The orientation of the orbital plane is specified by the longitude of the ascending node. 58.86004°, and the inclination i = 162.24220°.

When the numbers are located, look at the symbols used in the website, which may not be the usual ones because of font limitations. Since i > 90°, the motion of the comet in its orbit appears to be retrograde, opposite to the direction of motion of the planets, if the inclination is taken to be 180° – 162.24220° = 17.75780°.

Finally, the orientation of the orbit in its plane is given by the argument of perihelion, 111.8656°, measured from the ascending node in the direction of movement. It is not easy to comprehend the orbital position from the bare numbers; a drawing, made with some care, will show exactly what is going on.

As the perihelion distance is q = a(1 – e), it is a simple matter to determine a. In fact, a = q/(1 – e) = 0.58710374/ 0.03272276 = 17.941755 AU. From this, the orbital period P can be found. Since $P^2/a^3 = 1$ if P is in years (earth = 1) and a is in AU (earth = 1), P = 75.99716 years, very close to 76.0 years. The aphelion distance is 2(17.941755) – 0.58710374 = 35.296 AU.

This is outside the orbit of Neptune (a = 30) but not as far as Pluto (a = 39.5). In accordance with the law of areas, Halley's comet spends most of its time drifting in this dark region, periodically darting in to the sun to have a little more of itself boiled off into space.

Although orbital elements are given to high precision, the accuracy is not necessarily as high. Especially for comets, they can be changed by planetary perturbations, sometimes by large amounts.

The ejection of gases from comets also leads to reaction forces that can affect the orbit. Also, the reference coordinates may change with time, because of precession of the equinoxes and other effects, and this must be taken into account in accurate calculations. These changes are not changes in the orbit, of course.

DETERMINATION OF ORBITS

Finding the orbital elements from observations, and predicting the changes in orbital elements due to perturbations, are two of the most important problems in celestial mechanics, and have received close attention from Newton's time onwards.

We cannot give any reasonable account of this work here, but we can show how orbital elements come from observed motions by a graphical analysis that is very instructive, though of little practical use where high precision is required. The reader with drawing supplies is encouraged to follow along.

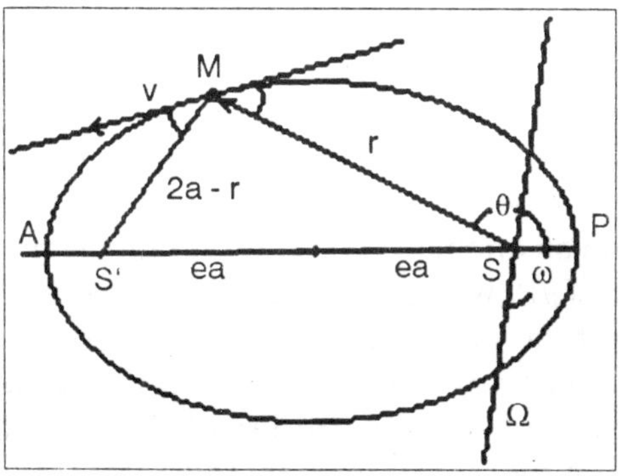

Fig. Determination of Orbital Elements

We suppose known the position r of a body M from the sun and its velocity v relative to the sun. These are six parametres that will serve to determine the six orbial elements. The orbital plane is defined by the plane of r and v, and can be represented in two views by means of orthographic projection. The line of intersection of the orbital plane with the ecliptic plane can then be found, and the dihedral angle between the two planes. Since the line of intersection will be the line of nodes, we have now found two of the orbital elements, Ω and i, which determine the plane of the orbit.

The next step is to calculate the mean distance a, using the energy relation $v^2 = GM(1/r - 1/2a)$. From a we can find the orbital period by $P^2/a^3 = 4\pi^2/GM$. The radius vector SM and the line from M to the other focus of the ellipse make equal angles with the tangent to the ellipse at M, which is in the direction of v.

In a view showing the orbital plane in true size, Lay off SM and MQ, where MQ is an arbitrary line in the direction of the empty focus. The empty focus S′ is located by laying off a distance 2a – r along MQ′, since the sum of the distances from the foci to M is 2a.

Now we can draw the line SS′; find its centre at O, and then the locations of perihelion and aphelion. The eccentricity of the orbit is e = OS/a = OS′/a. The orbit can now be drawn, and the angle from the line of nodes to the perihelion, the argument of perihelion, ω, can be measured, as well as the true anomaly θ, which is the angle PSM. We now have the five elements Ω, i, ω, e, and a.

The remaining element is the time of perihelion passage, which can be found from the true anomaly. First, we find the eccentric anomaly by $\tan(E/2) = [(1\text{-}e)/(1\text{+}e)]^{1/2} \tan(\theta/2)$. and from it the mean anomaly M = E – e sin E. The mean anomaly is $M = 2\pi(t - T)/P$, where T is the time of perihelion passage, and t is the time of observation. That is, the body passed through perihelion at a time $t - T = MP/2\pi$ earlier. We have now determined all six orbital elements from the six components of the initial position and velocity.

This was easy because the distance r was one of the given quantities. In general, orbits must be determined by observations of the angular position from earth, not directly in terms of distance and velocity.

The position at three different times is sufficient to

determine the orbit in this case, but the analysis is more difficult than what we have done.

Orbital elements vary only slowly in the solar system, except when there are near encounters, usually of light bodies with massive planets or satellites, that affect the orbits of the light bodies, but not of the heavier ones.

If we know the orbit at one instant, then we can predict the velocity and position at a later time from the orbit. We can also integrate the actual changes in position and velocity at the later time, taking into account forces exerted by bodies other than the sun.

The difference will be reflected in the orbital elements, which will slowly change. This is a useful way to take perturbations into account, and is widely used.

The maximum errors in using the Keplerian orbits are on the order of 5" to 30" for the inner planets, somewhat more for Jupiter and Saturn. Neptune was discovered in 1846 as a result of its perturbations of the motion of Uranus. This was a remarkable demonstration of the accuracy of Newtonian mechanics.

The earth's mean distance is decreasing by 5×10^{-8} AU per century, about 75 m per year. The eccentricity is decreasing by 3.804×10^{-5} per century. The inclination is decreasing by 46.94" per century.

These are average rates, and are simply the current values of changes that may be periodic. Nevertheless, they show how gradual any changes are, and allow the calculation of the earth's position with reasonable accuracy for thousands of years either side of the present. The longitude of the node, however, moves at -3.038" per *year* because of the precession of the equinoxes, which moves the reference point.

The argument of perihelion is increasing at 11.9828" per year. It is typical for an orbit to show a relatively rapid change in longitude of the node and motion of the perihelion; the moon is an excellent example, and also earth satellites, which are perturbed by the equatorial bulge of the earth and tidal forces exerted by the sun and moon, as well as by atmospheric drag if they are in low orbits.

MONGE'S PROCEDURE

Gaspard Monge, comte de Péluse (1746-1818) was a mathematician in the golden age of French mathematics, and a founder of the Ecole Polytechnique, the first modern engineering college.

He is known for the science called Descriptive Geometry, which is a method of representing objects in a two-dimensional drawing in such a way that all of their geometric properties can be investigated and described.

These methods were taught in classic form in every engineering course of study until computer graphics changed the approach.

The fundamentals are the basis of computer graphics and design, however. The classic methods were intended as much as training for the mind as means of calculation. In addition, they are inherently fascinating and enjoyable.

The first attempt to represent a three-dimensional object in a two-dimensional drawing is by means of some pictorial view, such as perspective, isometric, or oblique views. In any such attempt, some information is necessarily lost, and the object is only partially or ambiguously represented.

Antiquity used graphical methods almost exclusively in engineering calculations, a fact that was almost unrecognized, because of the perishability of such representations, and that the work was for practical, not literary, ends.

Here and there, survivals have given a glimpse into the extensive development of these methods, in architecture and shipbuilding. Among these methods was the use of two coordinated drawings to present three-dimensional information, as mentioned by Vitruvius.

Monge revived this science, extending it to the solution of many types of problems, and showing how it gives a complete solution of the problem. We will show a simple application here that will give an idea of how it is carried out. In the Figure, AB is a *plan view* of, say, a mine shaft. The x and y coordinates of points A and B are represented in this view. Line 1-1 is called a *folding line* that separates the view showing the same points as A' and B'.

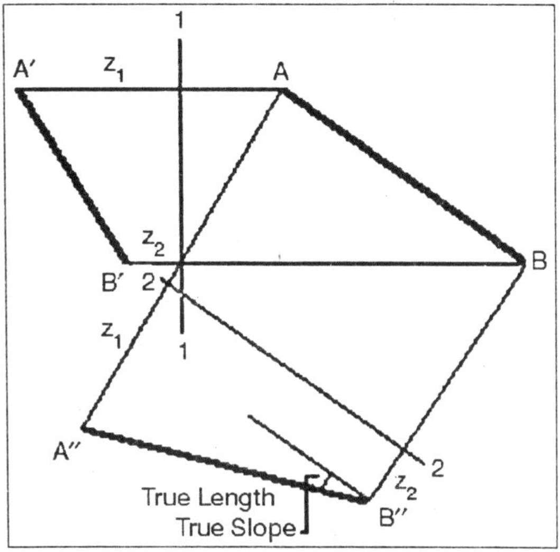

Fig: Monge's Procedure

This is a *side view* showing the coordinates y and z of the same two points. Note that the horizontal *projection lines* connecting A and A′, and B and B′, are perpendicular to the folding lines and show that the y-coordinates are the same. These two views give all three coordinates of any point in an object, and describe it completely. Of course, any two such views will suffice; we have made an arbitrary choice here. Suppose we want to know the length and slope of the tunnel between A and B. To find them graphically, we draw a folding line 2-2 parallel to the plan vies of the line.

The direction perpendicular to this line must represent the z-coordinate. Now we draw the two projection lines AA″ and BB″, and measure out from folding line 2-2 exactly the same distances from folding line 1-1 to A′ and B′. We have now located the projections of A and B in this view. The shaft lies in a plane parallel to this plane, so the true length and the true slope of the shaft are represented as shown, and can be scaled off.

We have shown the essential parts of Monge's Procedure, which are the selection of folding lines, and the measurement of equal distances in alternate views, by which any questions of size, shape, or orientation can be solved. This is the basis of orthogonal projection for engineering drawings, where three views are generally used to represent an object to make use of the redundancy for added vividness of presentation.

EARTH SATELLITES

The mass of an earth satellite is infinitesimally small compared to the mass of the earth, so the centre of the earth may be considered as a fixed point. A reference frame fixed to the earth's centre is not an inertial frame, due to revolution about the sun and the moon's motion. The earth-moon barycentre is a "more inertial" reference point, but in any case the difference from an inertial frame is negligible. There is a change in terminology, as perihelion becomes perigee and aphelion becomes apogee. This useless distinction is regrettable, and can be carried to excess (perijove, etc.). It would be good if there were terms usable in any orbit.

GM for the earth is $3.98600440 \times 10^{14}$ m^3/s^2. Therefore, $a^3/P^2 = 10097$ if a is in km and P is in seconds. For P = 86164.0989 s (sidereal rotation period of the earth), a = 42,164,172 m, or an altitude above the surface of 22,232 miles.

A satellite with this mean distance will return to the same point in the sky each day. If the inclination and the eccentricity are zero, the satellite will appear to hang motionless in the sky. Such an orbit is called *geostationary*, obviously of great value as a communications relay location. An area of very choice space real estate is thereby created, for which there is considerable demand. Satellites occupy fixed stations around the equator.

They remain in a constant direction from an observing point on the earth's surface, so an antenna can be permanently pointed at them without having to be continually redirected, which would not be easy nor cheap.

Their positions may be adjusted slightly to keep the satellites on station by on-board thrusters during their effective life, and the thrusters are used to remove them from orbit when their fuel runs out, since space is at a premium in the geostationary belt.

From a geostationary orbit, the earth subtends an angle of 17.4°, the same as a 12"-diametre globe at a viewing distance of 40", so it is quite reasonable to take photos of the whole visible hemisphere at once. Examples are available at the Dundee University website in the References. The poles are not visible, since the view extends only to latitude 81.3°.

The first artificial earth satellite was Sputnik I, launched on 4 October 1957 and weighing 83.6 kg. Its orbital period was 96.15 min, so its mean distance was 6953 km. The eccentricity was 0.0517, giving a perigee altitude of 228 km and an apogee altitude of 947 km. The inclination of the orbit was 64.26°, so it could easily be seen in every part of the earth when the sun shone on it.

Sputnik remained in orbit until January 1958, making 1350 revolutions. Because of the low perigee, it was considerably affected by atmospheric drag, its final period being near 90 min. The period of a satellite that just grazes the surface of the earth would be 84.48 min, and its velocity 7906 m/s.

The Global Positioning System (GPS) uses satellites that continually radiate very accurate timing information, corrected for relativistic time dilatation due to the satellite's speed, so that their distances from an observer can be found to within a few centimetres from the time differences.

The orbits must be known to a similar accuracy, so the effectiveness of this system is evidence of the correctness of the dynamics. Orbit information is broadcast along with timing information. The carrier frequencies are 1.57542 and 1.22760 GHz. The complete GPS system was deployed in 1993, consisting of 21 operational satellites and three spares, on circular orbits inclined at 55° and with a period of one-half a sidereal day (a sidereal day is about 23h 56m), so the satellites are seen to rise and set about 4 minutes earlier each day, and appear twice a day. Four satellites in good postions are intended to be visible from any point at any time, giving a good intersection, and eliminating the necessity for accurate calibration of the receiver clock (the offset of the receiver clock is one of the four unknowns that can be determined). The altitudes, about 20,200 km (a = 26,560 km), are high enough to make atmospheric drag negligible. There are six orbital planes, with four satellites spaced equally in each. The GPS is by no means the only satellite navigation system that has been developed, but it has become the most used, and is replacing the others.

Earth satellites are affected by many small forces in addition to the main inverse-square force directed toward the centre of the earth, and these *perturbations* cause the orbital elements to change slowly with time. In the case of earth satellites, the mean distance a, the eccentricity e and the inclination i do not change on the average (they may fluctuate slightly over the short term).

The longitude of the line of nodes, Ω, the argument of perigee, ω, and the rate of change of mean anomaly change steadily with time. The drag of the atmosphere is negligible for satellites that do not come lower than about 1000 km at perigee, which includes most practical satellites. For lower satellites, atmospheric drag causes a loss of energy at perigee that pulls in the apogee position, making the orbit less eccentric and decreasing the orbital period.

Paradoxically, the drag speeds up the satellite! The pulls of the moon and sun, *tidal effects* cause orbital changes. The direct tidal effect is due to the force exerted directly on the satellite, while the indirect tidal effect is due to the changes in mass distribution of the earth caused by tidal motions. Pressure of solar radiation and solar wind is another disturbing effect. The most important perturbing force, however, is that exerted by the equatorial bulge of the earth.

The gravitational potential of the earth can be expressed approximately as $V = GM/r\,[1 - (a_E/r)^2 J_2 P_2(\sin\varphi)]$, where a_E is the equatorial radius of the earth, 6 378 137 m, $J_2 = 0.001082630$, and $P_2(x)$ is the Legendre polynomial of second degree, $(1/2)(3x^2 - 1)$. The argument of this polynomial is usually seen as $\cos\theta$, but here we use the latitude $\varphi = 90° - \theta$.

This is the first part of an expansion in spherical harmonics, which can even allow for a lumpy earth that is not symmetric about the polar axis. The correction to the 1/r field shown here is a zonal harmonic of order 2, related to the flattening f = a/(a – b) = 1/298 of the earth. However, the gravitational potential depends on the distribution of mass in the earth, so it cannot be expressed simply in terms of f. The radial component of the force is $-\partial V/\partial r$, and the tangential part is $(1/r)\partial V/\partial\theta$. The motion of earth satellites has led to a much better knowledge of the earth's gravitational potential.

If a satellite has a mean motion $n = 2\pi/P$, a mean distance a and an inclination i, the movement of the line of nodes is $d\Omega/dt = -(3n/2)(a_E/a)^2[\cos i/(1 - e^2)^2]J_2$, and the change of the argument of perigee is $d\omega/dt = +(3n/4)(a_E/a)^2[(5\cos^2 i - 1)/(1 - e^2)^2]J_2$.

The quantities are in MKS units, and the results are in radians per second. For the moon, the line of nodes rotates once in 18.6 years. For a typical GPS satellite, the line of nodes moves about -0.03° per year, the perigee about 0.01° per year.

All the other perturbations are much smaller in amount, but must be taken into consideration in accurate work. Relativistic perturbing accelerations are inversely proportional to the fourth power of the distance r. Their magnitudes for GPS satellites are about 3×10^{-10} m/s^2, less than other small perturbations, so they have no practical effect, although they have been considered.

The International Space Station (ISS), shown at the right, is a famous earth satellite. The image is from the NASA website. The first two modules, Zarya and Unity, were assembled in orbit in 1998, and now there are 14. It is 73 m across the solar arrays, 44.5 m long and 27.5 m high, with 425 m^3 of habitable space. Power comes mainly from the 892 m^2 of solar arrays. There are thrusters to adjust the orbit when necessary. The total mass is about 179 metric tons. Although this is called "space travel∂ by NASA, it is not even outside the earth's atmosphere!

Approximate elements of the ISS orbit at 13.49Z, 6 April 2003, are $i = 51.6°$, Ω 19.99°, $e = 0.00083$, $\omega = 52.91°$, $M = 307.29°$, and mean motion 15.59579861 rev/day, or 0.0649825°/s. ("Z∂ is another designation for UT.) Exact orbits, predicted about 10 days in advance, can be found at the NASA link in the References.

The period is $P = 5539.95$ s (1.54 hr), and the mean distance is 6767009 m. This is a low, almost circular orbit at an altitude of about 400 km. Because the orbit is low, the ISS is seldom illuminated by the sun at night, so it is not frequently seen, though a frequent, large and prominent object.

The longitude (RA) of the ascending node changes rapidly, by about -5.00° per day, as does the argument of perihelion, by -3.64° per day. Perturbation by the oblateness of the earth is the reason for most of this change. The inclination, eccentricity and mean distance do not change rapidly. The orbit is just above the maximum ionization in the F layer of the ionosphere, and so is affected by atmospheric drag. The density of the atmosphere at this height is about $\rho = 9 \times 10^{-12}$ kg/m^3. The drag is given by $F = C\rho V^2 A/2$, where C is the drag coefficient, V the satellite velocity, and A the effective projected area. NASA gives $A = 344$ m^2 and $C = 2.36$. $V = 2\pi a/P = 6541$ m/s (assuming a circular orbit), so $F = 0.16$ N, which will produce an acceleration of 8.9×10^{-7} m/s^2. This drag causes the mean distance and period to decrease.

NASA notes that the "decay∂ is 4.11×10^{-4} rev/day^2. In 100 days, the mean motion will increase by 0.0411 rev/day, or $P = 5527$ s, a decrease of 13 s, or 0.23%. If uncorrected, the satellite would spiral inward at an increasing rate, eventually burning up catastrophically.

It may be interesting to find out where you should look for a satellite in a known orbit. The procedure will be illustrated here without taking all the refinements into account that are necessary for, say, GPS positioning.

The idea is to find the rectangular coordinates of the satellite in an approximately inertial system with its origin at the centre of the earth, and then to find the rectangular coordinates of the point of observation in the same system.

The differences of the coordinates then give a vector from the point of observation to the satellite. The motion of a satellite as seen from a fixed location on earth may be very complex, because of the interaction of the two motions involved, the revolution of the satellite and the rotation of the earth. Only when these are approximately equal is the situation more or less simple.

Let's suppose the satellite is in a circular orbit with a = 26 500 000 m and i = 55°, like a GPS satellite. Let the longitude of the ascending node be 40°. Since the orbit is circular, there is no perigee point, so we measure the true anomaly from the line of nodes in the orbital plane. Let the true anomaly at the time we are considering be 70°.

The z-axis is taken along the axis of rotation of the earth, and the x-axis in the direction of the vernal equinox. The y-axis then makes a right-handed coordinate system. This is just like the case of planetary motion, except that the role of the ecliptic plane is played by the equatorial plane. The orientation of these axes remains fixed in space as the earth revolves about the sun.

First, resolve the radius vector along and perpendicular to the line of nodes. The components are $a \cos \theta = 9\ 084\ 055.0$ m and $a \sin \theta = 24\ 958\ 236.0$. Then, $z = a \cos \theta \sin i = 20\ 444\ 590.0$ m, $x = a \cos \theta \cos \Omega - a \sin \theta \sin \Omega = -9\ 084\ 055.0$ m, and $y = a \cos \theta \sin \Omega + a \sin \theta \cos \Omega = 24\ 958\ 236.0$ m. This is the instantaneous position of the satellite in the inertial system, and it could obviously be found for any time equally easily.

Now for the point of observation. Let's use my own location, which is longitude $\lambda = -104.92583°$, latitude $\varphi = 39.72694$ and height above the ellipsoid (or mean sea level) h = 1633.7 m. Your own coordinates, if you do not know them already, can be found from a USGS topographic map. A meridional section of the earth is shown at the right, where φ is the geographic or spheroidal latitude.

The dimensions of the earth are a = 6378136 m and b = 6356752 m. There is a number of reference spheroids, any one of which will work satisfactorily. The eccentricity of the elliptical cross-section of the spheroid can then be found to be e = 0.006694167. The meridional radius of curvature at latitude φ is $N = a[\cos^2\varphi + (1 - e^2)\sin^2\varphi]^{-1/2}$, or 6 378 194.38 m at my location. The length of N below the equatorial plane is e^2N = 285.8 m.

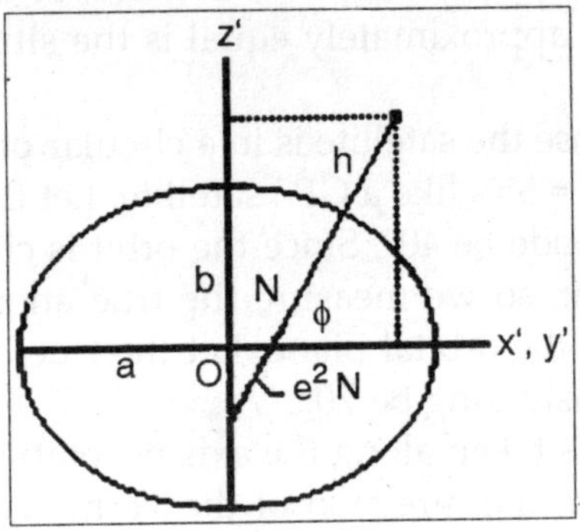

Fig: Geographic to Rectangular

We now choose an earth-fixed rectangular coordinate system with the z′-axis along the rotational axis of the earth, the x′-axis in the meridian of Greenwich, and the y′-axis making a right-handed system. The rectangular coordinates of my location are then $x' = (N + h)\cos\varphi\cos\lambda$ = -1 263 816.21 m, $y' = (N + h)\cos\varphi\sin\lambda$ = -4 741 167.76 m, $z' = (N + h - e^2N)\sin\varphi$ = 4 077 353.73 m. These coordinates do not change as the earth rotates. The distance from the centre is 6 379 711.32 m.

The relation between the inertial and earth-fixed coordinate systems is shown at the left. The earth-fixed system rotates steadily in the positive direction about the common z and z′ axes, and the angle of rotation Θ is the local apparent sidereal time at Greenwich, which is easy to find from the Astronomical Almanac.

The rotational period of the earth is 23h 56m 04.09890369732s. When the time is stated to such precision, there are many small corrections to be considered, but we shall ignore them here so the principle is not lost.

We shall assume that the inertial and earth-fixed coordinates are related simply by the rotation through an angle Θ, not correcting for polar motion, precession of the equinoxes and nutation. These are simply rotation matrices applied in addition to the one for the main rotation. For accurate work, however, they cannot be ignored.

Since sideral time goes from zero to 24 hours in one rotation, we must multiply an ordinary time interval by 24/ 23.9344719 or 1.00273781195 to find the equivalent sidereal time interval. A sidereal hour is simply equivalent to 15° of angle; it is not an "hour∂ in the usual sense of time.

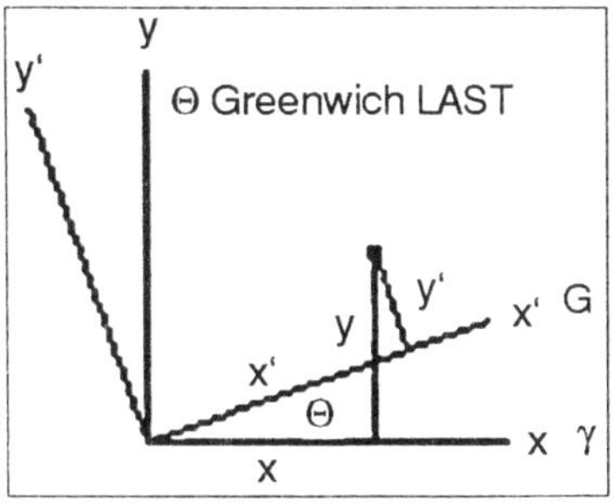

Fig: Inertial to Earth-Fixed Coordiantes

The transformation equations are the familiar $x = x' \cos \Theta - y' \sin \Theta$ and $y = x' \sin \Theta + y' \cos \Theta$. All you have to do is check that the signs of the sines are correct.

If we put in coordinates in the earth-fixed system (x′,y′,z′) we will get out coordinates in the inertial system (x,y,z) which we can compare with the satellite coordinates. Let's pick 10.00 am MST on 2003 January 1. Since my time zone is +7, the UT is 17.00h January 1. From the Astronomical Almanac, page B8, the sidereal time at 0h UT was 6.61651428 hours.

The sidereal time elapsed is then 1.00273781195 × 17.00 = 17.04654280 hours. Adding the two, the Greenwich LAST at my 10.00 am will be 23.66305708 hours, or 354.945856°. If it seems easier, this can also be expressed as –5.0541438°.

The results are x = –1 676 585.46 m, y = –4 611 395.05, z = 4 077 353.73 m. The square root of the sum of the squares is 6 379 711.32, so the arit7hmetic checks.

Now we can subtract the coordinates of the observation point from the coordinates of the satellite to find the relative vector (X,Y,Z). We find X = –7 407 469.54 m, Y = 29 569 631.05 m, Z = 16 367 236.27 m. The direction of this vector can now be expressed in terms of right ascension and declination, since it is a vector in the inertial system referred to the vernal equinox.

The distance from the observer to the satellite is 34 599 423.53 m. The projection on the equatorial plane is 30 483 334.55 m, so that the declination $\delta = 28.2324°$ and the right ascension is 75.9363° or 5h 3m 45s. At the date and time specified, this direction is beneath the earth. In fact, the antipodeal point of right ascension 18h and declination -29° is in the southern sky at an altitude of about 20° above the horizon.

Although these calculations are tedious and subject to error, even when done with an electronic calculator, a computer programme can do them with ease, speed and correctness, and even the more precise calculations are not much bother. It would not be difficult to write a programme to determine the visibility of all of the 24 GPS satellites, and the directions in which they are to be seen.

OTHER ORBIT LORE

The total energy E of a body of unit mass moving under the influence of a potential energy per unit mass V(r) is the kinetic energy $T = v^2/2$ plus V(r). In polar coordinates, the components of velocity are dr/dt and r $d\theta/dt$, so the total energy becomes $E = (dr/dt)^2/2 + r^2(d\theta/dt)^2/2 + V(r)$. The constancy of angular momentum for a central force gives $r(d\theta/dt)^2 = h = \text{constant}$.

Substituting for $d\theta/dt$, we find $E = v^2/2 + h^2/2r^2 + V(r)$. This is the energy equation for a particle of unit mass moving in one dimension with the *effective* potential $h^2/2r^2 + V(r)$. The additional term is called the *centrifugal potential*. In the case of orbital motion under gravity, $V(r) = -GM/r$. The form of the resulting potential is shown in the figure. For $E = 0$, a particle coming in from infinity accelerates steadily until $r = h^2/GM$, then is rapidly decelerated, coming to rest at $r = h^2/2GM$, and afterwards retracing its path in reverse. Of course, the particle swings round the centre of force in this motion. For $E = -(GM)^2/2h^2$, r has the constant value that gives the minimum of the curve, and the orbit is circular. For intermediate energies, r oscillates between a minimum and a maximum value, corresponding to perhelion and aphelion.

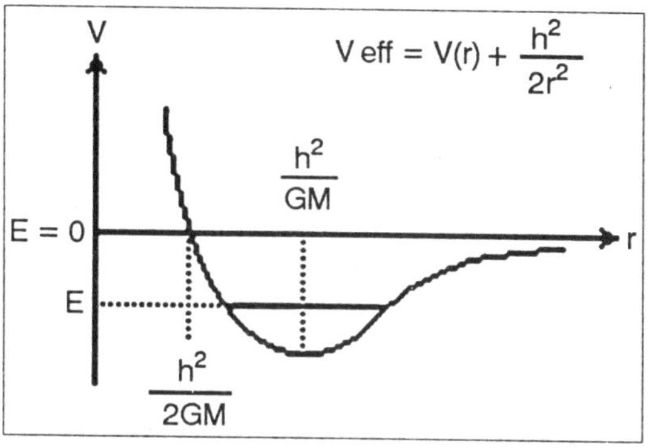

Fig: Effective Radial Potential

Consider the vector B = v × h + (GM/r)r. The angular momentum per unit mass, h, is here taken to be a vector perpendicular to the orbital plane. The cross product then lies in the orbital plane, and so then does B. Taking its time derivative, we have (dv/dt) × h – $(GM/r^3)r(v\cdot r)$ + (GM/r)v. Now dv/dt = $-(GM/r^3)r$, and h = r × v, so the first term becomes - $(GM/r^3)[r(r\cdot v) - r^2v]$. All the terms cancel, so dB/dt = 0, and B is a *constant of the motion*. This is extraordinary, since we already have one constant of the motion, the total energy E, and this is usually the only one that can be found.

For an inverse-square force, we have found a second one. It is known that this happens when we have an additional symmetry in the problem—here it happens to be symmetry under rotation in a four-dimensional space, but there is no room to explain this here..

Since B is a constant, it can be evaluated at any point of the motion, and it is easiest to choose perihelion.

At this point, v × h = $v_p^2 r_p$, so B is a vector parallel to the line between the focus and perihelion. We will find the same vector if we evaluate B at any point of the motion. Its constancy is a good proof of an exact inverse-square force.

We have seen that with an attractive inverse-square force, closed orbits exist for negative energies. For zero or positive energies, the orbits are open. We have considered briefly the zero-energy parabolic orbits that divide the two classes, but not the hyperbolic orbits, since they are not of much use in the solar system.

For a repulsive inverse-square force, there are no closed orbits of negative energy at all, only the hyperbolic orbits of positive energy. On these orbits, the body comes in from infinity along a path asymptotic to a line of a certain direction.

It then interacts with the other body, in what is called a "collision,∂ though the bodies do not come into contact, and leaves to go to infinity asymptotic to a different direction.

MOTION WITH CONSTANT ACCELERATION

The usual case of motion with constant acceleration is motion under the influence of gravity, an elementary problem of great importance that is treated in every introductory Physics course. The equation of the hodograph is then dv/dt = g, which integrates to the straight line gt, whose length is proportional to the time of the motion. If the initial velocity v_o is drawn from the origin to a point P, then the hodograph is a vertical line lengthening downwards.

The velocity at any time is easily read off, and if the motion ends at a point Q, the corresponding velocity is $v_f = v_o + gt$, making a triangle OPQ with side PQ vertical. The trajectory, of course, is a parabolic arc that is more difficult to work with than the simple triangle. These things are illustrated in the diagram at the right.

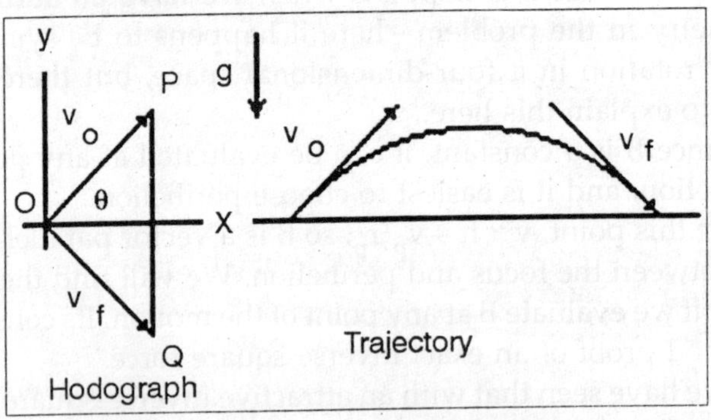

Fig: Projectile Motion

The area of the triangle is half the base, gt, times the altitude $v_0 \cos\theta$, where θ is the angle between the initial velocity and the horizontal, the angle of projection.

Also, gt/2 is $v_0 \sin\theta$. Eliminating v_0, we find the altitude to be $gt/(2 \tan\theta)$, and its area $(gt/2)^2/\tan\theta$. This area is proportional to the horizontal distance covered by the trajectory, since the altitude is the (constant) horizontal component of the initial velocity, and the base is proportional to the time. The proportionality constant of the area of the triangle to the horizontal distance covered is g/2, of course.

To find the condition for the maximum range as a function of the angle of projection, differentiate the area with respect to θ and set the derivative equal to zero.

This gives $\cos 2\theta = 0$, or $2\theta = 90°$, or $\theta = 45°$, the well-known result. Therefore, the initial and final velocities are perpendicular, so that the hodograph is an isosceles triangle.

For a particle projected directly upwards, the hodograph will be a vertical line of length 2v with its centre at the origin. The time of flight t is given by 2v = gt, or t = 2v/g. At time t/2, we are at the origin, and the velocity is zero.

The area of the triangle is zero, meaning that the particle goes nowhere horizontally, as we know. The hodograph of a particle moving with constant velocity is a point, and the corresponding trajectory is a straight line. These simple relations are easy to visualize, and help to illustrate the meaning of the hodograph.

The hodograph of a uniform circular motion with angular velocity ω and radius r is a circle of radius $v = r\omega$, the peripheral velocity, with its centre at the origin.

The hodograph is a quarter-revolution ahead of the trajectory, which is also a circle, of radius r. The circular hodograph implies steady rotation. These relations are shown in the figure at the right. The constant magnitude of the acceleration is $v^2/r = \omega^2 r$, directed toward the centre of the trajectory circle and perpendicular to v. Note that this *centripetal* acceleration is in the same direction in the hodograph and the trajectory. The radii of the circles differ by a factor of ω.

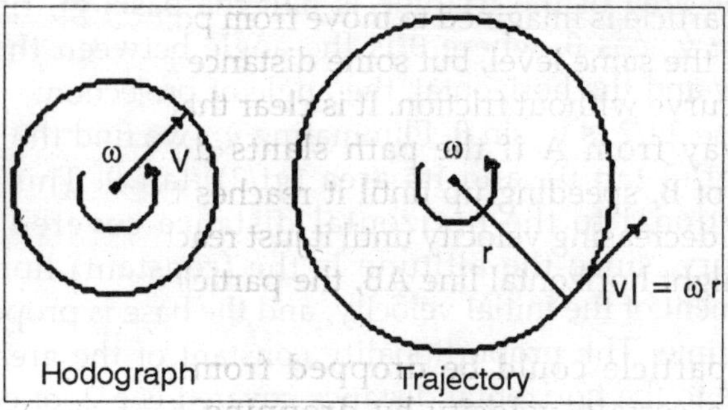

Fig: Uniform Circular Motion

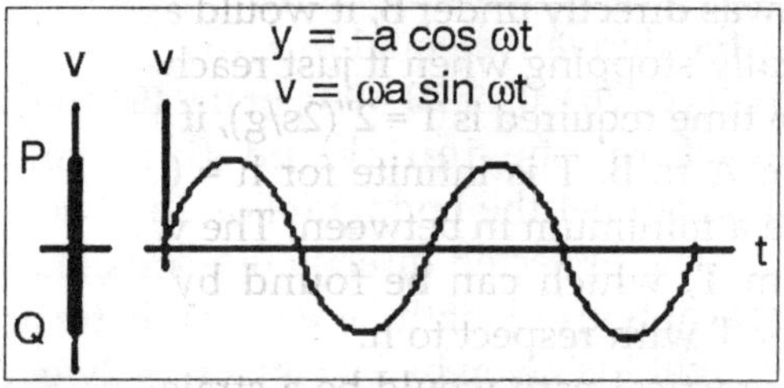

Fig: Hodograph: Simple Harmonic Motion

The hodograph of a simple harmonic motion is shown at the left, with the time dependence displayed. The amplitude of the motion is a and its angular frequency is $\omega = 2\pi/T = 2\pi f$, where T is the period and f is the frequency.

The velocity also varies sinusoidally, with an amplitude ωa. Points P and Q correspond to the passage of the particle through y = 0, and the centre of PQ to the extreme positions.

THE BRACHISTOCHRONE

This is a famous problem of constrained motion under gravity. The problem was published by Johann Bernoulli (1667-1748) in 1696, and was solved by him, Isaac Newton (1642-1727) and G. W. von Leibniz (1646-1716) shortly afterwards.

It was the stimulus for the development of the calculus of variations by Euler and Lagrange, in which an unknown function is determined that minimizes a definite integral. The name is well-chosen from Greek, from "brachistos,∂ shortest, and "chronos,∂ time.

The particle is imagined to move from point A to a second point B at the same level, but some distance away, sliding on a certain curve without friction. It is clear that the particle will move away from A if the path slants downward in the direction of B, speeding up until it reaches the lowest point, then with decreasing velocity until it just reaches B. If the path is the straight horizontal line AB, the particle will never start moving.

The particle could be dropped from A, and when it attained a certain velocity by dropping a distance h, be deflected horizontally in the direction of B.

When it was directly under B, it would again be deflected upwards, finally stopping when it just reaches B. In this case, the minimum time required is T = 2"(2s/g), if s is the horizontal distance from A to B. T is infinite for $h = 0$ or for $h = \infty$, so there must be a minimum in between. The value $h = s/4$ gives the minimum T, which can be found by minimizing the expression for T with respect to h.

A more practical path would be a straight inclined ramp for half the distance on each end. The effective acceleration on a ramp making an angle θ with the horizontal is $g \sin \theta$, so it is easy to calculate the total time in this case as well.

The answer is the same, 2"(2s/g), for a depth at the centre of $h = s/2$ (a 45° ramp). Suppose we were using this for a rapid transit system, and two stations were 1000 m apart. In this case, $\Phi(2s/g) = 14.3$ s, so the total time would be 28.6 s, and the average speed 35 m/s or 142 km/h or 89 mph.

This would be rapid transit indeed, with the maximum speed no less than 223 mph (the speed attained by falling 500 m). Friction, however, stands in the way of this energy-efficient transport system. Proposals have been made to use gravity for this purpose, but no bold ones have been realised.

If the path is given by the function $y = y(x)$, then the speed at any point x is $v = \Phi(2gy)$, and the path length $ds = \Phi(1 + y'^2)$ dx, where $y' = dy/dx$, so the time required to go from 0 to x can be expressed as the definite integral $T = \int(0,x)\ [\Phi(1 + y'^2)/\partial(2gy)\ dx$. We search for the path y(x) that gives a minimum value for the integral.

What sort of path would give us the absolute minimum time? We can anticipate that it would be a smooth path, replacing the linear ramps and abrupt falls and rises, and that the minimum depth would be between s/4 and s/2.

Bernoulli, Newton and Leibniz found the surprising result that the path is a cycloid, and the maximum depth is s/π. The time required is $\Phi\pi\partial(2s/g)$, so $\Phi\pi = 1.7725$ replaces 2. The time for our rapid transit is then 25.3 s instead of 28.6 s.

This path is the *brachistochrone,* which gives a shorter time than any neighbouring path.

It is clear that since the time is a minimum for the brachistochrone, the time will not be greatly different for any approximate path. Even the ramps give only a 13% difference.

A parametric equation for the cycloidal path is $x = (s/2\pi)(\theta - \sin\theta)$, $y = -(s/2\pi)(1 - \cos\theta)$. The angle θ is the angle of rotation of a circle of radius s/2π rolling on the underside of the x-axis, and goes from 0 to 2π. This curve is shown at the right, with the generating circle indicated. It is also the path of an electron starting from rest at A in an electric field in the direction of -x, and a magnetic field out of the plane, in the z-direction. It is not easy to find this result with the usual methods of the calculus of variations because of algebraic complexity. However, the hodograph offers an easier solution.

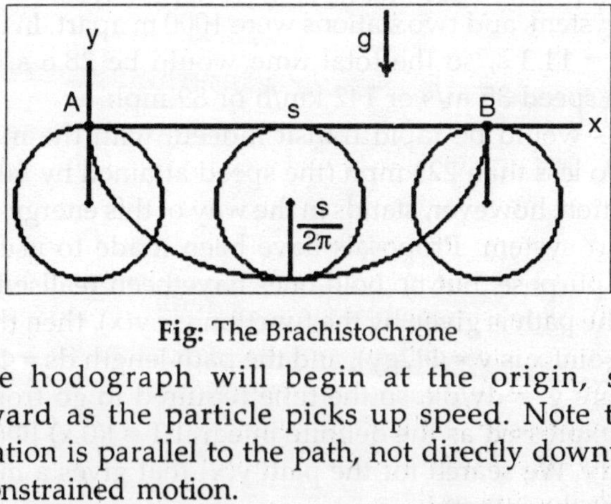

Fig: The Brachistochrone

The hodograph will begin at the origin, sloping downward as the particle picks up speed. Note that the acceleration is parallel to the path, not directly downward as in unconstrained motion.

The hodograph will then loop around, reaching the

horizontal axis at the minimum of the curve, where the velocity will be greatest, then looping around above as velocity decreases, and finally approaching the origin with a downward slope.

Integrals can be written for the area of the hodograph (time of the motion), as well as for the horizontal distance covered, in terms of a hodograph u(θ). [This is not the parametre θ of the preceding paragraph, but the inclination of the velocity vector.] The area of the hodograph will be proportional to the time of the motion, so we look for a curve that will minimize the area inside the hodograph, while covering the necessary horizontal distance s.

The first integral is to be minimized, with the second as a condition, for which the method of the Lagrange multiplier is convenient. Apostulatos shows that the hodograph will be a circle tangent to the vertical axis at the origin. It is, in fact, $v(\theta) = \Phi(2gs/\pi) \cos \theta$. The coefficient of the cosine is the diametre of the circular hodograph.

If we add a constant velocity to the left so that the hodograph becomes a circle centred at the origin, that is, a velocity $\Phi(gs/2\pi)$, we get the hodograph of a uniform circular motion. Therefore, the actual motion is a circular motion translating to the right at a speed equal to the peripheral speed of the circular motion. This is, of course, a cycloid. The total time is then the distance s covered by the rolling circle divided by its peripheral velocity, $\Phi(gs/2\pi)$, which gives the expression quoted above.

PLANETARY ORBITS

The first two of Kepler's Laws of Planetary Motion are: (1) the planet moves in an ellipse with the sun at one focus; and (2) the area swept out by the radius vector is the same in equal time intervals.

An elliptical orbit is shown at the right, in which the planet is located by its radius vector r and the true anomaly θ. The velocity is greatest at perihelion and least at aphelion, and in both cases is perpendicular to the radius, which attains a minimum and maximum value, respectively. In between, there is a component of the velocity dr/dt parallel to the radius. The quantity p is the semi-latus-rectum, and the mean distance, or semimajor axis of the ellipse, is a, given by $p = a(1 - e^2)$, where e is the eccentricity, and $0 \le e < 1$. The focus is displaced by a distance c = ae from the centre of the ellipse.

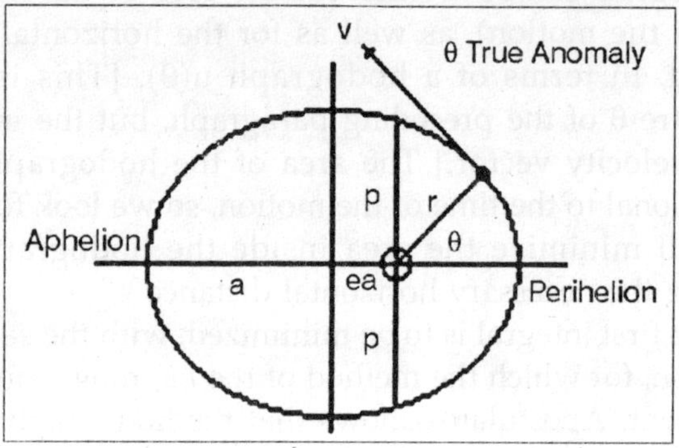

Fig: Trajectory

The polar equation of the orbit is shown at the left, together with the mathematical expression of the Law of Areas. The Law of Areas is equivalent to the conservation of angular momentum, which is proportional to the constant h.

These two expressions are derived directly from Kepler's Laws. If we differentiate r with respect to time, we can find the radial and transverse components of the velocity. The parametre θ is the true anomaly, and the radial and transverse components are relative to this direction. Plotting these components will give us the hodograph of the motion.

$$r = \frac{p}{1 + e\cos\theta} \qquad p = a(1 - e^2)$$

$r^2\dot{\theta} = h$ Law of Areas

$v_r = \dot{r} = (he/p)\sin\theta$

$v_\theta r\dot{\theta} = (h/p)(1 + e\cos\theta)$

The hodograph is shown at the right. It is, remarkably, a circle with centre at O, which does not coincide with the origin unless e = 0 and the orbit is a circle.

To draw the hodograph, all that is necessary are the

velocities at perihelion and aphelion. These are laid off above and below the velocity origin, and form the diametre of the hodograph. The velocities marked 90° and 270° are the velocities at the ends of the latus rectum. Note that they fall at the intersections of a horizontal line through O with the hodograph.

If we add a downward velocity of he/p, the centre O is brought to the origin, and the hodograph of this motion is a uniform circular motion. This fact can be used to solve the problem of the position in orbit as a function of time (Kepler's Problem), and the angle of rotation is the mean anomaly, which increases uniformly with time.

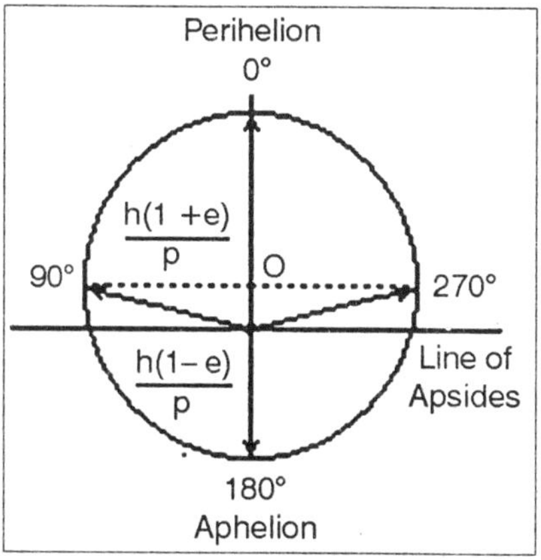

Fig: Hodograph of an Orbit

Bringing in Kepler's Third Law, that the cubes of the mean distances are proportional to the squares of the orbital periods, allows the determination of the law of force, which turns out to be the familar inverse square.

To obtain this result, note that the acceleration in the circular motion is the total acceleration, since the velocity he/p is a constant, and it is easy to find the force for a uniform circular motion.

The reader is encouraged to solve the problem of a circular orbit under an inverse-square force, and from it to show that Kepler's Third Law results. Then, it is not difficult to see how to prove the inverse, which is the desired conclusion.

ELASTIC COLLISIONS IN THE SOLAR SYSTEM

The article by Van Allen quoted in the Reference recalled my attention to the subject of encounters between orbiting bodies and the interesting effects on their orbits.

Apparently, this is a subject of some mystery to physics students, though it is really quite simple in principle. The reader is strongly encouraged to read Dr Van Allen's excellent article, in which the encounter of Pioneer 10 with Jupiter in December 1973 is discussed as an example. The article, despite its clarity, still has the aura of mystery, or at least complicated physics, here and there.

One source says that the first use of gravitational assist was with the Mariner 10 probe of 1973-1975.

An encounter with Venus on 10 March 1974 was used to decelerate the vehicle and send it towards Mercury. Mariner 10 made three passes around Mercury, giving us most of the information on the surface of that planet that we have to date. This date is later than the December, 1973 date for the Pioneer 10 encounter with Jupiter, so perhaps Pioneer 10 was actually the first.

Jupiter is about 5 AU (the AU is the radius of the earth's orbit) from the sun, and revolves at an orbital speed of about 13.1 km/s on the average, completing its orbit in 11.857 years. Pioneer 10 approached Jupiter on an elliptical orbit at a speed of 9.8 km/s, not far from aphelion. After its closest approach of 2.027×10^5 km, it swung away, approximately tangent to Jupiter's orbit, at a speed of 22.4 km/s, which was sufficient to put it into hyperbolic orbit about the Sun. Since then, Pioneer 10 has left the solar system, and just recently its final radio transmission was received.

This is an example of *gravitational assist*, which has been frequently used in planetary probe trajectories, performing feats that would be impossible, or very costly, if done directly.

Although we do deal with three bodies here, the sun, Jupiter and the spacecraft, none of the complicated results of the celebrated "problem of three bodies∂ is required for an understanding. There is a massive body, Jupiter, and a body of negligible mass, the spacecraft, in independent orbits about the even more massive Sun. A close encounter between Jupiter and the spacecraft, which is an elastic collision in which the bodies interact gravitationally, takes place in a short time and limited space to change the instantaneous velocities of the participants, sending them into new orbits.

The new orbit of Jupiter is indistinguishable from the old one; though slightly different, the difference is extremely small. The new orbit of the spacecraft, however, can be very different, and its kinetic energy can change by a considerable fraction. This energy comes from Jupiter, of course, which does not notice the theft. In the elastic collision, the total kinetic energy and total momentum of the two bodies is conserved. However, it may be distributed differently before and after the collision, as is very well known. Let us reduce the problem to its simplest state, that of a one-dimensional collision, as on an air track. An actual collision in three dimensions may approach this more or less closely when the impact parametre (distance of closest approach if the bodies did not interact) is small.

First, let's assume that the large mass M is moving with velocity w to the right, and the small mass m is moving with velocity v to the left, as shown in the figure. The magnitudes of the vectors are shown above the arrows giving the directions. The relative velocity is v + w, and in the elastic collision is reversed in the centre of mass system in which the centre of gravity is fixed. This system moves with velocity w to the right, very closely, since M >> m.

Therefore, after the collision, the mass m is moving to the right with a velocity v + 2w. In a head-on collision, the light body can pick up twice the velocity of the heavy body. In the figure, momentum and energy are apparently not conserved, but this is only because the very small decrease in w is not shown. An accurate figure would differ imperceptibly.

Now suppose the large mass is still moving with velocity w, but the small mass comes up behind it with velocity v > w, approaching with relative velocity v – w. In the centre of mass system this velocity is reversed after the collision, and the final velocity of the small mass is v – 2w. In a rear-end collision, the small mass can lose twice the velocity of the heavy body.

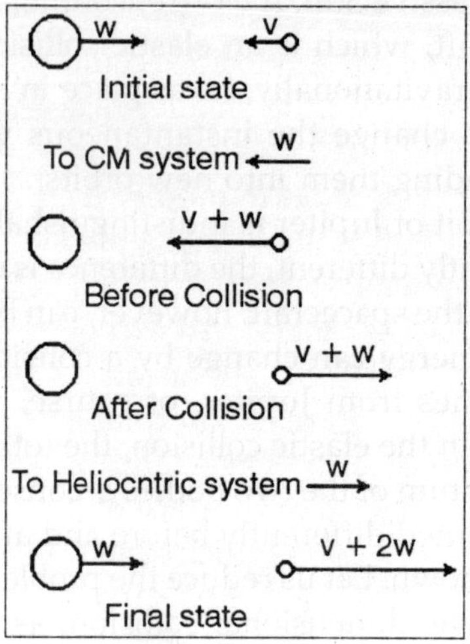

Fig: Head-On Collision

We have now found the limits of the velocity of the light body after collision, and how these limits depend on the velocity of the heavy body. In a three-dimensional collision, the velocity change will be smaller, and can easily be calculated explicitly for any case. In the case of Jupiter, the speed of Pioneer 10 after the encounter could be as high as 9.8 + 26.2 = 36 km/s, or as low as zero. The actual result of 22.4 km/s is well within the possible range. An unlucky encounter could send the spacecraft plummeting rapidly into the Sun.

Comets have long been known to interact with Jupiter in this way. If they come in behind Jupiter, they lose energy and are put into short-period elliptical orbits, many of which do indeed come perilously close to the sun. If they come in ahead, they are flung well out of the solar system, like Pioneer 10. Saturn exerts a very similar effect. Saturn's orbital velocity is 9.67 km/s, so the changes it can make are not as large as those of Jupiter. If Pioneer 10 got a kick in the apse from Saturn, it could only with difficulty attain the required escape velocity. Uranus and Neptune move even more slowly, but still could perturb a parabolic comet into an elliptical one. These massive planets can create a tight orbit in an encounter that leads to orbit modification.

Mars, with 24.1 km/s; Earth, with 29.8 km/s; and Venus, with 35.0 km/s orbital velocities could transfer large velocities in an encounter, but the smaller masses do not reach as far. Any planet is sufficiently massive relative to a comet, spacecraft or meteoroid to modify its orbit with no change to its own. It should be noted that a planet cannot *capture* a body in this way, in a two-body encounter. It is the velocity relative to the Sun that is changed. An exception is when the collision is not elastic, and the body encounters some atmosphere or even the solid surface of the planet. This is, of course, of frequent occurrence on Earth, and has occurred elsewhere as well, as the pock-marked surface of the Moon attests. We also note that encounters in which kinetic energy is lost generally give rather elliptical orbits; circular orbits are not to be expected.

GRAVITY, DENSITY, FORCE AND MASS

Newton's Law of Gravitation does give a certain force directly, the gravitational force $f = Gmm'/r^2$ between two masses. If m and m′ are in gram, r in cm, then $G = 6.672 \times 10^{-8}$ dy-cm^2/g^2. The universal gravitation constant G is not known to great precision because it is equivalent to weighing the earth—not a convenient thing. It, too, gives only a ratio of force to mass, since $f/m = Gm'/r^2$.

If r is the radius of the earth, 6378 km, and m′ its mass, 5.98×10^{27} g, then f/m at its surface is g = 981 cm/s^2, called the acceleration due to gravity. This is a convenient standard acceleration for us to use to relate force and mass, where now the force is called *weight*, to imply the standard acceleration. Actually, we measure g and use it to determine the mass of the earth, of course.

A problem with this is that g varies with position. At any one point, it is constant enough, but has different values at different points, which gives the same mass different weights at different places.

The g we observe is the joint result of the gravitational attraction of the earth and the centrifugal acceleration due to its rotation. At the equator, g is about 978.039 gal ("gal" stands for Galileo, and is used in gravitational studies for cm/s^2, together with the smaller milligal = 10^{-3} gal). Here we are furthest from the centre of the earth, and the centrifugal force is greatest. At the pole, g = 983.217 gal. The difference is about 0.5%.

We are closest to the centre of the earth, and the centrifugal force is zero. At any latitude φ, $g = 978.049(1 + 0.0052884 \sin^2 \varphi + 5.9 \times 10^{-6} \sin^2 2\varphi)$ gal. Here, Heiskanen's 1928 value for the equatorial gravity is used. If we are above sea level, the *free-air* correction 0.3086h mgal must be subtracted, where h is the altitude in metres.

There are other corrections depending on the masses above and around the points, and when all this has been considered, there are still "anomalies" of up to about 100 mgal that cannot be explained by anything obvious. All this is to show that the actual acceleration of gravity is not a uniform value. The value g = 981 cm/s^2 is a good value for everyday use.

Since we know the mass and the radius of the earth, we may itch to know its average *density*, or ratio of mass to volume. Its density $d = m/(4\pi r^2/3) = 5.5$ g/cm^3. The ratio of the mass of a body to the mass of an equal volume of water is called its *specific gravity*, sp.gr., and in the cgs system is numerically equal to the density, but is dimensionless (and so the same in any system of units).

The specific gravity of granite is 2.64-2.76, of dolomite 2.84, of limestone 2.68-2.76, of sandstone 2.14-2.36, of feldspar 2.55-2.75 and of mica 2.6-3.2. A good characteristic value for surface rocks is about 2.7. The earth can't be made up of surface rocks! Hematite, an iron ore, has a sp.gr. of 4.9-5.3.

Rock salt, halite, is 2.18, glass 2.4-2.8, bituminous coal 1.2-1.5, crude oil 0.8, oak 0.7 and pine 0.5. The sp.gr. of aluminum is 2.8, zinc 7.1, cast iron 7.2, steel 7.9, brass 8.5, copper 8.9 and lead 11. The sp.gr. of ice is 0.917, so it floats.

The sp.gr. of water is 1.00 at 3.98°C, 0.99823 at 20°C, and 0.95838 at 100°C. Mercury, used in barometres and such, has 13.5953 at 0°C and 13.5462 at 20°C. Dry air has a density of 1.2929 g/liter at 0°C, and its density at any other temperature and pressure can be estimated from the gas laws. These are representative values that show the range for common materials, and can be used in approximate calculations.

If we must relate mass and force, however, some such value must be assumed as a constant conversion factor.

It is no longer the actual acceleration of gravity, but an arbitrary constant. Nevertheless, it should be close to an average value of g. The value 980.665 gal has been chosen to fill this role. It would have been just as easy to have chosen 981 gal, but no, we have to make it look like something precise. When we use this value, try to eliminate any suggestion in your mind that it involves some kind of physical measurement. It is just an arbitrary conversion constant.

THE FPS SYSTEM

The absolute foot-pound-second system of units (fps) retains the second as the unit of time. The unit of length is the foot, 30.48 cm (exactly). Whoever decided that an inch should be exactly 2.54 cm has my thanks. Another definition, 1 metre = 39.37 in, which gives 1 inch = 2.54000508 cm, was used to define the *US survey foot,* used in geodesy. There is no difference practically, only an annoyance in measurements of high precision.

The third fundamental unit is the pound, the unit of *force.* The Handbook of Chemistry and Physics tells that 1 pound = 0.45359237 kg. Now kg is mass, so this really means that 1 pound force = 1 lbf = (0.45359237)(980.665) dyne = 4.4482216×10^5 dynes = 4.4482216 N, or, that a pound force is the weight of 453.59237 g.

Someone went to a lot of trouble to compare the standard kilogram in Paris with the standard pound in London, and came up with all those digits for us to admire.

A choice of 1 lb = 454 g would have done just as well to define the fps system, with only a negligible amount of restandardization. The standard acceleration of gravity in the fps system is 980.665/30.48 = 32.17405 ft/s^2. In everyday problems, the value 32.2 ft/s^2 is normally used. The foot-pound is 1.35582 J.

It is usual to refer to the weight of a body in the fps system, not its mass. One cc of water has a mass of 1 g, and so a weight of 1/453.59237 lb in fps. A cubic foot of water is 28316.8 cc, so it weighs 62.428 lb at 3.98°C or 62.317 lb at 20°C.

The figure of 62.4 is commonly accepted by engineers. Using the specific gravity, we can now find the weight of a cubic foot of anything. Since w/m = g, m = w/g in the fps system, so we have f = (w/g)dv/dt for Newton's Law. The appearance of g caused the name "gravitational system" to be applied to this kind of fps system, but it is a misnomer. There is nothing gravitational about it, and it does not depend in any way on the acceleration of gravity.

The "g" is only a conversion constant that happens to be approximately equal to the acceleration of gravity. It is *never* anything other than 980.665. The unit of mass has been named the *slug*. The weight of a slug is about 32.17 lb. It is equal to 14.593 kg. Mass densities should be specified as slug/ft^3.

Civil engineers adopted the slug, mechanical engineers didn't appreciate the problem, and chemical engineers did something else which shall mention in a moment. The symbol γ is used for weight density, γ = w/V in, for example, pounds per cubic foot, pcf. Mass density is still ρ or d.

There are two general ways to determine the weight or mass of a body. One way is to use a *balance*, where the mass is compared with a known mass using levers or the equivalent. This includes the once-common laboratory beam balance.

The result is unaffected by variations in gravity. The other way is with a device that measures the force directly, such as a spring scale or strain gauge. Such a *spring scale* must be calibrated, and the calibration will depend on the local value of gravity.

Household spring scales are not sensitive enough to detect variations in gravity. Most small "balances" are now electronic spring scales.

There are other ways of defining fps systems. One way lets the unit of mass be the pound, or lbm. Then f = mdv/dt defines the unit of force, called a *poundal*, pdl. The force acting on 1 lbm in standard gravity is, then, 32.2 pdl, which must be its weight. 1 pdl = 13,810 dynes, we can easily discover. This system was used in school physics for many years, but never really caught on except with physics teachers. Another way keeps both the pound mass (lbm) and the pound force (lbf), which means that Newton's law has to be written

$$f = (m/g)dv/dt,$$

causing all kinds of mischief. Note that this is not just,

$$f = (w/g)dv/dt,$$

which is harmless, since w/g is just mass. By contrast, m/g isn't anything holy. Here, g is not just a conversion factor, but is an essential part of the equation, contributing units as well as magnitude. Chemical engineers used this system.

Since they don't use Newton's Second Law much, it doesn't do a lot of harm, and allows the carefree confusion of two different units with the same name.

In using the fps system, it is best to use feet and pounds consistently, and to divide them decimally, just like metres and kilograms. If it is not otherwise obvious, whether factors of g should be introduced can be determined from the units. For convenience, things like inches, miles, tons and acres can be brought in later, but not used in theoretical calculations.

For example, standard atmospheric pressure is 2117 psf, and water weighs 62.3 pcf, air 0.075 pcf. A speed of 30 mph is 44 fps. Quantities can be expressed in consistent units by factors such as these before performing a calculation.

It is a subsidiary matter that the cgs system generally uses the Celsius thermometre, and the Kelvin for absolute temperature,

$$K = °C + 273.15.$$

The fps system generally uses the Fahrenheit thermometre and degrees Rankine for absolute temperature,

$$°R = °F + 491.69.$$

Of course 5K= 9°R.

Either temperature scale can be used with either system of units. Specific heats are numerically the same in cal/K and Btu/°R. 778 ft-lb = 1 Btu, and 4.186 J = 1 cal.

The gas constant is 8.31441×10^7 erg/K-gmol, or 1545.33 ft-lb/°R-lbmol.

INTEGRALS OF THE EQUATIONS OF MOTION AND CONSERVATION LAWS

Let's now look at integrals of the equations of motion. If we multiply by dt and integrate, we find,

$$\begin{aligned}\int f dt &= \int m(dv/dt)dt \\ &= m\int dv \\ &= mv_2 - mv_2 \\ &= \Delta mv.\end{aligned}$$

$\int f dt$ is called *impulse*, and mv is the *momentum*. Neither of these quantities has any special name, but the units are ML/T: gm-cm/s, dyne-s or lb-s.

Multiplying by dx instead,

$$\begin{aligned}\int f dx &= \int (dv/dt)dx \\ &= \int (dv/dx)(dx/dt) \\ &= \int v dv \\ &= mv_2^2 - mv_1^2 \\ &= \Delta(mv^2/2).\end{aligned}$$

These quantities are much used: $\int f dx$ is *work*, and $mv^2/2$ is *kinetic energy* T. In cgs, the unit of mechanical energy is the erg (from the Greek for work), the dyne-cm, and in fps, it is the ft-lb. The MKS unit is joule = N-m, so $1\ J = 10^5 \times 10^2 = 10^7$ erg. The time rate of change of kinetic energy is $dT/dt = mv(dv/dt) = fv$, called *power*. It can be expressed as erg/s or ft-lb/s, and 10^7 erg/s = 1 W. An erg/s is 0.1 μW. In fps, we may use the colourful horsepower, 1 hp = 550 ft-lb/s. Since 1 ft-lb = 1.356×10^7 erg, 1 hp = 745.7 W.

If f is a function of position, f = f(x,y,z), and $\int f dx$ around any closed curve is zero (as it is in the gravitational field of the earth), then it a function V(x,y.z) can be found such that $f_x = -\partial V/\partial x$, or, in vectors, f = -grad V. This procedure is very familiar from electrostatics. The function V is the *potential energy*.

Then,

$$\begin{aligned}\int f dx &= -\int (dV/dx)dx \\ &= V_1 - V_2 = -\Delta V.\end{aligned}$$

Combining this with our earlier expression, we have $\Delta V + \Delta T = \Delta(T + V) = 0$, or $T + V = E =$ constant. This is the principle of *conservation of mechanical energy*, which proves very useful.

Energy leads the way into advanced dynamics, where it replaces the concept of force. For example, the potential energy function $V = kx^2/2$ yields the force $f = -kx$, which we know gives us the harmonic oscillator. We use Newton's Law to give $-kx = md^2x/dt^2$, or $x'' + (k/m)x = 0$, where the prime indicates differentiation with respect to t.

The solution of this equation is $x = A \sin \omega t + B \cos \omega t$, with A and B constants that are determined by the initial conditions, and $\omega = \Phi(k/m)$. In advanced dynamics, we form the Lagrangian $L = T - V$ and use Lagrange's equations to get the equations of motion. This is only added complication in this simple case, but the method is capable of great generalization.

Another important conservation law comes from Newton's Third Law, which states that if one body exerts a force on another, then the second body exerts an equal and opposite force on the other. This says that if

$$f = dp_2/dt,$$

then $-f = dp_1/dt$, from which it follows that

$$d(p_1 + p_2)/dt = 0,$$

so that $p_1 + p_2$ = constant.

This is the principle of the *conservation of momentum*. Each component of the momentum is separately conserved. The conservation of energy and momentum are powerful tools for the study of collisions. An *elastic* collision is one in which the energy is conserved. Even in an inelastic collision, the momentum is still conserved.

If the force is constant, then we can integrate the equation of motion twice and find the position as a function of time. With $(f/m) = a$, we find $x = at^2/2 + v_0t + x_0$ with two constants, the initial velocity and position. This solution is studied in every elementary physics course, and is applicable to the everyday problem of projectile motion under gravity. There are many useful consequences of this solution, such as,

$$v^2 - v_0^2 = 2a(x - x_0),$$

where the time has been eliminated between it and its first derivative. In the form $v = \Phi 2as$ it is very useful, even in things like hydrodynamics.

DYNAMICS IN POLAR COORDINATES

So far we have not emphasized the vector nature of Newton's Second Law, since in rectangular coordinates we can write one equation for each dimension and work entirely with components. In fact, many practical problems are one-dimensional, and the subject does not come up.

When we consider motion in a plane described by polar coordinates, however, we must look more deeply.

If a point is given by P(r,θ), then the unit vector r′ is r/r, and the unit vector θ′ is at right angles to it toward the direction of increasing θ, as shown in the diagram at the right. The primes will have to do here for identifying the unit vectors.

Now, since,

$$\begin{aligned} r &= rr', \\ v &= dr/dt \\ &= (dr/dt)r' + r(dr'/dt) \\ &= (dr/dt)r' + r\omega\theta', \text{ since } dr'/dt = \omega\theta', \end{aligned}$$

where ω is dθ/dt. Note also that dθ′ = –ωr′, because we need this for the acceleration: $a = [d^2r/dt^2 - r\omega^2]\, r' + [2(dr/dt)\omega + r(d^2\theta/dt^2)]\theta'$. This expression has many applications.

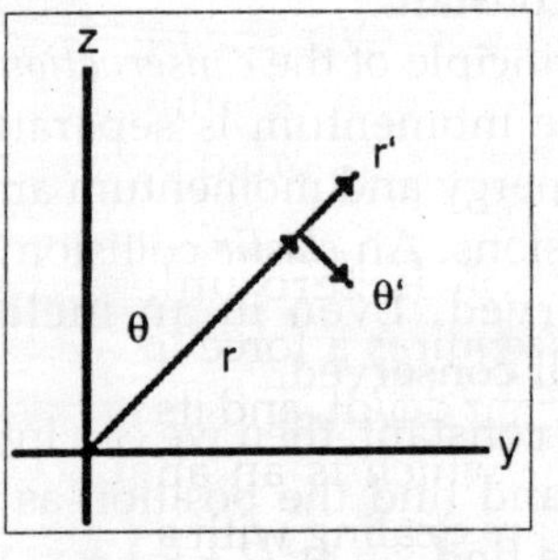

Fig: Polar Unit Vectors

Suppose first that the force acting on a particle moving in a plane is *central,* that is, directed toward the origin. Then the θ component of the acceleration must vanish, or 2(dr/dt)ω + r(dω/dt) = 0. If this is multiplied by r, we find that $d(r^2\omega)/dt = 0$, or $r^2\omega = h$, a constant.

Then, $mr^2\omega = mv_pr = H$, a constant. Here, v_p is the component of velocity perpendicular to the radius.

In vector notation, H = mr × v is called the *angular momentum*, the moment of the linear momentum about the origin. With a central force, then, the angular momentum is a constant. One consequence of this is Kepler's Area Law.

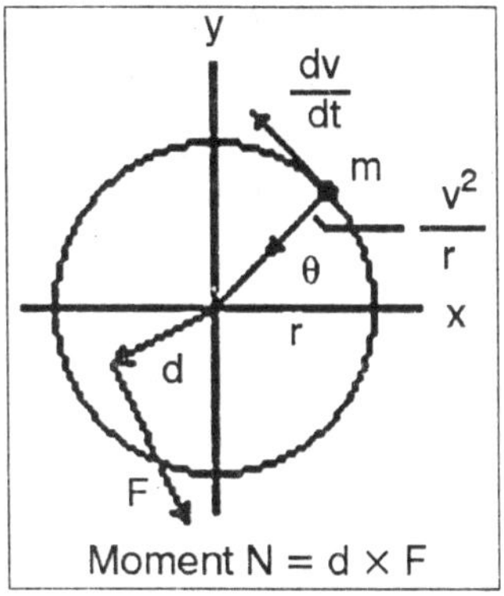

Fig: Circular Motion

Now suppose the particle is moving in a plane with constant r, as shown in the figure. This is the case for all the points in a rigid body rotating about fixed axis normal to the plane.

Then $a = -r\omega^2 r' + r d\omega/dt\theta'$. The first term is an acceleration directed toward the origin of amount v^2/r, the *centripetal acceleration* that must be supported by a force directed toward the orgin.

The second term is zero unless there is an *angular acceleration*, which requires a force in the direction of dv/dt.

This force is f = mr dω/dt, and its moment about the origin is $N = mr^2\ d^2\theta/dt^2$, which is an angular form of Newton's Second Law. If we are dealing with a rigid body, then $I = \Sigma mr^2$ is called the *moment of inertia of mass.*

We can then write $N = I\ d^2\theta/dt^2$, with N analogous to force. In the diagram, the applied moment N balances the accelerating moment. The forces do not have to have the same direction or magnitude, just the same moment.

In vectors, if f = dp/dt, then N = r × f = r × (dp/dt) = (d/dt)(r × p) = dH/dt, since v × p = 0 because v and p = mv are in the same direction. This shows that N = dH/dt in general.

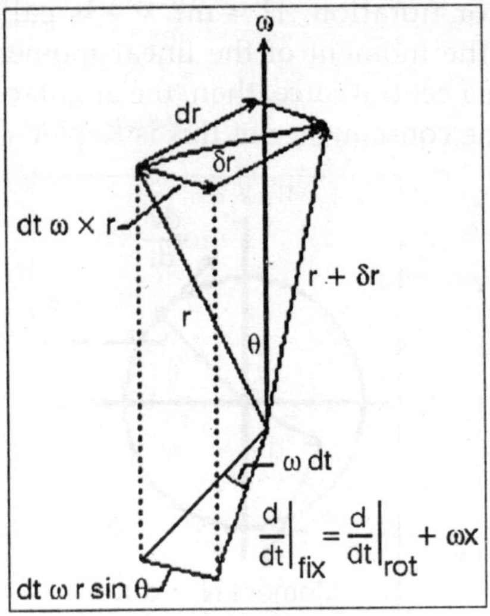

Fig: Time Derivative in a Rotating System

INERTIAL SYSTEMS, ROTATING SYSTEMS AND THE CORIOLIS FORCE

So far we have only considered one coordinate system, in which Newton's Second Law holds. For this reason, it is called an *inertial system*. Newton's First Law defines what is meant by an inertial system in which the Second Law can be applied. Consider a second system, moving with uniform velocity u with respect to this system. If primes refer to this system, then r' = r + ut. Then, the velocity and acceleration in the moving system are v' = v + u and a' = a. Since the acceleration is the same, Newton's Law holds as well in the moving system.

Systems moving with constant relative velocities to an inertial system are also inertial. This is the *principle of Galilean relativity*, which is valid so long as the relative velocities are much smaller than the speed of light. At greater velocities, we must use Einstein's principle of relativity instead.

Now let's consider a system rotating with a uniform angular velocity ω with respect to an inertial system.

The diagram shows that in addition to the change in a vector in the rotating system, dr, the change ω × r must be added to get the full change δr in the inertial system. This is a general result for the time derivatives of a vector, that can be applied to the velocity as well. We have $v' = v - \omega \times v$, and then $a' = a - \omega \times (\omega \times r) - 2\omega \times v$. Obviously, the rotating system is not inertial.

However, it will seem like an inertial system if we add two *fictitious* forces to the acceleration, as defined by the second and third terms. The first is the *centrifugal force,* the second the *Coriolis* force.

The centrifugal force is independent of the velocity and proportional to the square of the angular velocity, and is in a plane normal to the axis of rotation. The Coriolis force is proportional to the angular velocity and the linear velocity, and is normal to the plane defined by them.

A coordinate system fixed to the surface of the earth is such a rotating system. The angular velocity $\omega = 2\pi/86\,164.1$ s $= 7.29211 \times 10^{-5}$ radian/s. The sidereal (with respect to a fixed direction in space) period is used, not the period relative to the sun that we use for time. This is a small value, but has noticeable effects.

As we mentioned above, the centrifugal force is included in the effective gravitational force mg, so the only fictitious force to be considered is the Coriolis force. The surface of the earth is normal to the direction of the effective gravity.

Centrifugal force is the reason the earth is a spheroid. The effective component of the angular velocity is normal to a horizontal plane at the point of observation, with magnitude $\omega \sin \varphi$, where φ is the latitude. Its positive direction is upward in the Northern hemisphere, downward in the Southern hemisphere.

The direction of the Coriolis force is normal to a horizontal velocity, so that it would turn the velocity clockwise in the Northern hemisphere, and anticlockwise in the Southern.

The Coriolis force has little effect on the water draining out of a bathtub, but a great effect on large-scale winds, where it causes winds to blow along isobars, instead of directly into a low-pressure area or out of a high-pressure area. The result is that winds circle a low-pressure area anticlockwise and a high-pressure area clockwise in the Northern hemisphere and that these areas persist as they move about. This is certainly the most visible effect of the Coriolis force, and additional proof that the earth rotates relative to an inertial system.

The Coriolis force also causes the plane of a pendulum to rotate clockwise in the Northern hemisphere, as can be seen by considering the velocity of the bob.

The rate of this rotation is proportional to sin φ, so it vanishes at the equator and is a maximum at the poles. At the poles, the angular velocity is equal to that of the earth. At my latitude, 40°N, the pendulum makes one rotation in about 37 hours and 20 minutes. Foucault first observed this rotation, in 1851, in a famous experiment, and it is displayed in every good Science Museum. It is very difficult to create an unbiased support for the pendulum, and to release it properly, so it is a difficult home experiment. A ball dropped from a height in the Northern hemisphere winds up to the East of the point from which it was dropped (for a 200 ft drop, the distance is 0.31 inches).

FRICTION AND VISCOSITY

Coulomb not only verified the inverse-square force between electric charges and measured them absolutely, but also established the laws of friction. Friction is an extremely variable phenomenon, but one of considerable practical interest to engineers. Coulomb's main result was that the force preventing two solids with plane faces in contact from sliding was independent of the area of contact, but proportional to the normal force pressing the two solids together. The ratio of the frictional force to the normal force is called the *coefficient of friction*, $\mu = F/N$. Moreover, when the two solids were sliding, the coefficient was independent of the velocity of sliding. However, the *static* coefficient was somewhat greater than the *kinetic* coefficient. Frictional forces that obey these rules are called Coulomb friction. The coefficient of friction is dimensionless, the same in every system of units.

The coefficient of friction depends greatly on the preparation of the surfaces in contact. Clean, smooth surfaces will sometimes seize together in a vacuum so that they cannot be easily separated. It is never a good idea to have the same metal in contact at a bearing. For these reason, measurements of the coefficient of friction are not of great use, except to show average values. Nevertheless, we should have some idea of the range of values. Dry wood on wood shows $\mu = 0.25 - 0.50$, wood on metal 0.2 – 0.6. Steel on steel is 0.58 dry, but falls to 0.16 when lubricated with light oil. Steel on brass, lubricated, gives 0.19.

The viscosity of a liquid or gas expresses the force necessary to move surfaces with respect to one another with the liquid or gas between them. It is assumed that the fluid adheres to the metal on either side, establishing a velocity gradient, as shown in the diagram.

The viscosity is the ratio of the force per unit area in the direction of motion, divided by the velocity gradient, or ç = (F/A)/(v/d) = Fd/Av.

In cgs, the units are dyne-s/cm^2, given the name of poise, after Poiseuille. In fps, they are lb-s/ft^2. Using the known unit ratios dyne/lb and ft/cm, it is easy to find that 1 lb-s/ft^2 = 478.8 poise. In fps, the lb-s/in^2 or Reyn is often used, which is 3.325 poise. The viscosity of water at 20°C is 0.01 poise, or 1 centipoise, easy to remember.

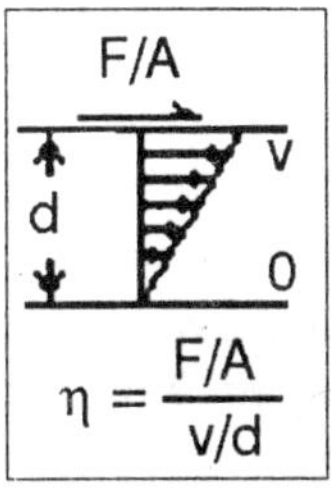

Fig: Coefficient of Viscosity

The viscosity of liquids is a strong function of temperature. In water, it varies from 1.787 cp at freezing to 0.2818 cp at boiling. Glycerine has 1490 cp at 20°C, hexane (0.6603) 0.326 cp and methanol (0.7914) 0.597 cp. Light machine oil has 34.2 cp at 100°F (37.8°C). These figures give an idea of the viscosity of common liquids at normal temperatures.

The viscosity divided by the density is found in enough formulas that it has a name, *kinematic viscosity*, and a unit, *stokes*, of its own. Since the density of water is 1, its kinematic viscosity in stokes equals its viscosity in poise.

The densities given above allow you to find the kinematic viscosity of a few liquids. Air, at 18°C, has a viscosity of 182.7 μpoise, which shows a typical value for a gas viscosity. Since its density is much lower than that of a liquid, its kinematic viscosity is not much different from that of a liquid (about 18 cp). Hagen and Poiseuille studied the flow of water and blood, respectively, in capillary tubes around 1839-40. The flow through a tube of radius r and length s with a pressure difference of p dyne/cm^2 between its ends is $q = \pi pr^4/8sç$ cm^3/s, where ç is the viscosity in poise. The *Reynolds number* R for this flow is $R = DV\rho/ç$, where D is the diametre of the pipe, and V the velocity of flow. Poiseuille's formula is valid when R is less than about 2000. Above this value, the flow becomes turbulent and different rules apply.

Another famous case of viscous flow is the movement of a sphere of radius a through a viscous fluid.

In this case, the force resisting the motion is $F = 6\pi çaV$ dynes, where ç is in poise, a in cm, and V in cm/s. A glass marble 13 mm in diametre, falling in glycerine, will have a terminal velocity of $V = (4\pi/3)(0.65)^2 (981)(1.24)/6\pi(14.90) = 7.67$ cm/s. This is found by equating the viscous force to the weight of the marble (sp.gr. = 2.5) less the buoyant force of the glycerine (sp.gr. 1.26). The Reynolds number in this flow is $R = DV\rho/ç$ as with Poiseuille flow, and in this case is only 0.84, so viscous flow is guaranteed. Millikan used the same theory in his oil drop experiment to determine the charge on the electron. He measured the weight of a drop by the rate at which it fell in air.

Examples

Let's solve the pendulum with a small amplitude from what we have found above. This is motion in a plane with constant r, and the force on the bob of mass m is downward, and of amount mg. The moment of this force about the origin is N = mgl sin θ, tending to decrease θ, so the equation of motion is mgl sin θ = $-ml^2 d^2\theta/dt^2$. Note that the tension in the cord has no moment about the origin, and so does not enter into the equation of motion.

For small oscillations, sin θ H″ θ, so we have θ″ + (g/l)θ = 0, where the primes indicate differentiation with respect to time. The solutions of this differential equation are well known, and θ oscillates sinusoidally with angular frequency Φ(g/l). A seconds pendulum, then, has a length $l = g/\pi^2$ = 99.4 cm. The period of a seconds pendulum is 2 seconds, one second for each swing that releases the escapement one notch, so the clock ticks once a second. The pendulum was once used to measure g, but now the fall of a ball is measured electronically, to an accuracy of a milligal or better.

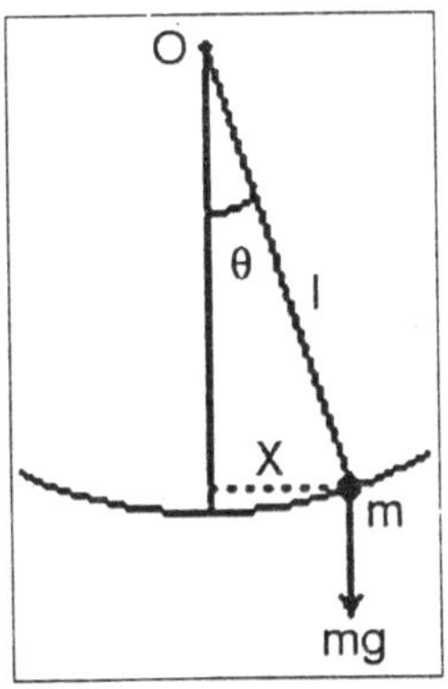

Fig: Pendulum

A ball leaves a cannon with a speed v_0 of 1800 fps and an elevation of θ = 15°. How far will it go on a level field? The initial upward velocity is 466 fps, so the ball will reach the summit of its trajectory at a height h in 466/32.17 = 14.48 s, and fall back to earth in 28.97 s.

In this time, it will travel horizontally d = 1739 fps x 28.97 s = 50,362 ft, or 9.54 miles. Of course, it will fall well short of this, because air resistance has a large effect, especially on supersonic cannonballs. If the muzzle velocity is only 180 fps, then it will travel for only 2.897 s, and plow up the ground only 504 ft away. This would be closer to correct, but it would be a really puny charge. Note that the range is not proportional to the muzzle velocity, but to its square.

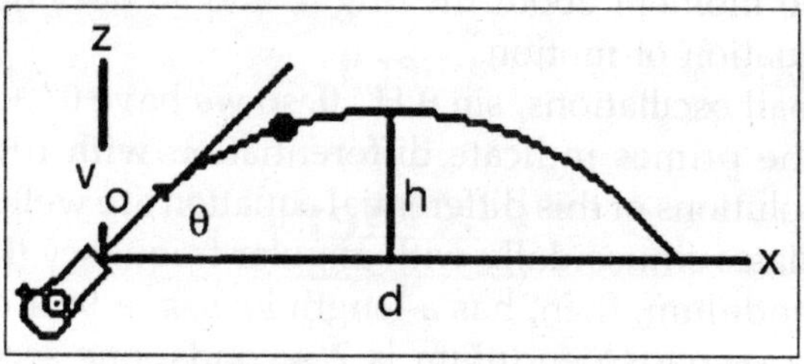

Fig: Trajectory

Such simple problems as these could be multiplied indefinitely. They are not only excellent exercises, but also build confidence in your ability to understand at least this much of the universe.

Physical Constants In CGS

- Speed of light c = 2.9979×10^{10} cm/s
- Electron charge e = 4.80325×10^{-10} esu
- Electron mass m = $0.9109534 \times 10^{-27}$ g
- Planck's constant $h/2\pi = 1.05459 \times 10^{-27}$ erg-s
- Bohr magneton $\mu_B = 0.9274078 \times 10^{-20}$ erg/gauss
- Boltzmann's constant k = 1.38066×10^{-16} erg/K
- Universal gas constant R = 8.31441×10^{7} erg/K-mol

Chapter 9

Property Kinematics

DYNAMICS

Let's begin by stating the formulas that are the goal of this analysis. They deal with the movement of viscous fluids, commonly air and water, and are some of the rare analytical results of fluid mechanics of practical value. We will study the dynamics of viscous fluids in this article, and show how these results can be obtained by analysis.

Poiseuille was interested in the flow of a fluid, actually blood, in narrow tubes, and found that the discharge Q in cc/s was related to the pressure difference Δp in dynes/cm^2 by $Q = (\pi a^4/8L\mu)\,\Delta p$. Here, a is the radius of the tube, and L its length, both in cm, and μ is the absolute viscosity of the liquid in dyne-s/cm^2 or poise.

The viscosity of water is 0.0101 poise at 20°C. Blood, with which Poiseuille was concerned, is not a simple Newtonian fluid, and its flow is a complicated problem, but water, oil and such obey the equation very well. The catch is that the flow must be rather slow, or the flow will become turbulent, obeying different rules and the resistance to flow will increase. The flow is guaranteed to be laminar (not turbulent) provided that the Reynolds number $R = DV\rho/\mu$ is below 2000, as found from experiment. Here, D is the diametre of the tube, 2a (cm), V is the average velocity of flow $Q/\pi a^2$ (cm/s), μ the absolute viscosity, and ρ the density in gm/cc. This number is dimensionless, and is the same in any consistent system of units.

For a 1/2" pipe, R is 2000 for an average velocity of 15.7 cm/s of water, or Q = 20 cc/s, which is not a very rapid flow.

We should also be aware that the laminar flow is not established immediately at the entrance to the tube, but takes a little length for this to happen, in which the liquid at the centre speeds up, and that at the wall of the pipe is retarded.

Stokes solved the problem of the steady motion of a sphere with velocity V through a still viscous fluid (or the flow of the fluid past a stationary sphere, which is the same thing). He found that the force exerted on the sphere by the fluid was $F = 6\pi\mu aV$, where F is in dyne, a in cm, V in cm/s, and μ is the absolute viscosity.

Note that the force is proportional to the velocity, the rule we often assume for a frictional force. Again, the motion must not be too fast, and a Reynolds number $R = DV\mu/\rho$ can be used to decide what is too fast.

This has the same form as in Poiseuille's equation, but the flow is different, and the critical value may be different. If V is too great, again the flow becomes turbulent, with the formation of a turbulent wake behind the sphere, and now the force is proportional to the square of the velocity. We usually write $F = C\,A(\rho V^2/2)$, where C is the dimensionless *drag coefficient*. Where Stokes's formula applies, $C = 24/R$.

The critcal Reynolds number is about 1 for air, and the drag coefficient for high R is about 0.4, dropping to about 0.2 when the boundary layer changes from laminar to turbulent. This is a complex and interesting subject that you can read more about in the references.

The streamlines in Stokes flow dip in toward the sphere, so to avoid disturbing the flow pattern, there must be no nearby boundaries. A ball dropping in a tube about twice its diametre would be significantly affected (the drag would increase).

Its area of application is really to rather small spheres in air, such as cloud particles, or to slow motion in a viscous fluid, such as glycerin or oil. It was famously used by R. A. Millikan in the determination of the electronic charge, to measure the weight of the drops by how fast they fell.

The drops were so small that a correction for the mean free path of air molecules was necessary for an accurate result.

The problem can also be solved neglecting viscosity, and the results are entirely different. Some are applicable to reality, but others are not. When a sphere moves in an ideal fluid, it suffers no drag whatsoever, but once started in motion, would continue forever. The effective mass of the sphere is increased by half the mass of the fluid displaced, a result that is quite valid and observable.

It can be applied to the motion of bubbles in beer, for example. Most real-life problems of fluid flow depend greatly on viscosity, even when it would seem to have a small effect. The drag on a body moving in a fluid is mostly caused by the lower pressure in the turbulent wake behind it, and the drag is reduced by making this wake as small as possible.

The pressure on the front of the body has only a small effect. Trucks with the "streamlining" in front of the trailer would do better to put it behind!

We recall that the coefficient of viscosity μ is defined by Newton's equation $F/A = \mu(dv/dy)$, where F is the shearing force on an area A, and dv/dy is the velocity gradient normal to the surface. If μ is a constant independent of the velocity of shearing, the fluid is called *Newtonian*. Many, if not most, fluids met with in engineering are Newtonian. Some, like paint and blood, are not, and there is a variety of interesting behaviours in some other substances, especially those that are a mixture of phases. μ is measured in *poise*, dyne-s/cm^2.

STRESS AND STRAIN

To begin our study, we must review the nature of stress and strain. This will require the use of quantities with a number of components in three-dimensional space, such as vectors, as well as more complex quantities. These can be represented explicitly, and were indeed for many years, but this involves tedious repetitive writing.

Vector analysis provides an excellent shorthand, and is widely understood and used. For more complex quantities, this was extended to what were called *dyadics*, but these proved rather cumbersome, and seem to have disappeared.

What turns out to be very convenient is representation

by Euclidean tensors using indices. This essentially writes all the components, as was originally done, but in a very compact and useful form.

2The vector a becomes the quantity a_i, where we understand that the index i takes the values 1, 2 and 3, or x, y and z, or whatever. The scalar produt of a and b is $a_i b_i$, where the repeated index implies summation over its range of values, called *contraction*.

Suppose that x_i is a point in space occupied by a solid body, and that $\hat{\imath}_i$ is the displacement of that point when the body is strained. Then, $\varepsilon_{ij} = \partial_i \hat{\imath}_j$ is the *strain*, or ratio of displacement to distance.

The symbol ∂_i is short for ∂/∂_i. This is a new kind of quantity, depending on two directions. Suppose this region of the body isn't really strained, just rotated bodily by an angle θ. Considering the displacements that result from such a rotation, it is clear that $\varepsilon_{ij} = -\varepsilon_{ji} = \theta$. We really don't want these if we are studying strains, and can get rid of them by forming $\Phi_{ij} = (\varepsilon_{ij} + \varepsilon_{ji})/2$, which is called *symmetrizing* Φ. What's left is just the *pure strain*, described by the *symmetric tensor* $\Phi_{ij} = \Phi_{ji}$.

There are six independent components of this tensor, three diagonal and three off- diagonal.

Among other things, this means that the order of the indices is immaterial, so we do not have to worry about it. In a fluid, instead of the displacements, we use the rate of displacement, the time derivative of $\hat{\imath}_i$ instead (which is just the fluid velocity v), so Φ becomes the *rate of strain* tensor.

Strain causes stress, so we must find some way to specify stress. If we imagine a plane cutting the solid, then the partA of the solid on one side of the plane exerts a force on the other side. The ratio of this force to the area considered is the *stress*.

Given a normal unit vector to specify the orientation of the plane, then the stress is another vector that is a function of the direction of the normal vector, as well as of position.

We expect, then, that stress can be described by a quantity Ψ_{ij}. By considering the equilibrium of a small volume, the fact that it must not rotate is reflected in the *symmetry* of Ψ. Like Φ, then, Ψ is symmetric and has six independent components.

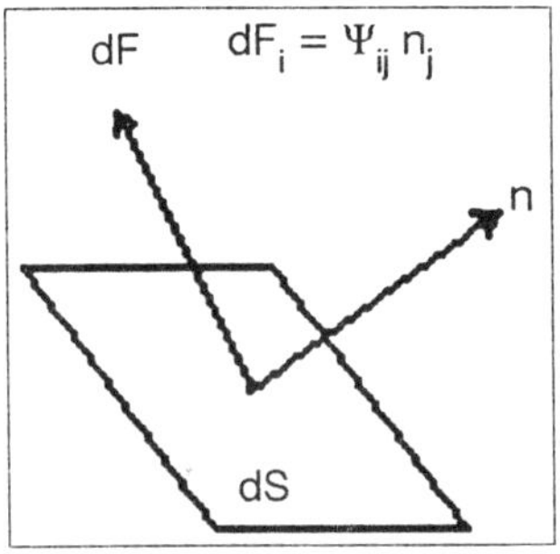

Fig: Stress Tensor

In a fluid at rest, the only stress that can exist is a pressure normal to any area, and this pressure is the same in any direction. This is, in fact, the definition of a fluid.

Therefore, considering areas normal to the coordinate axes, we find that $-pn_j = \Psi_{ij}n_j$, where n_i is a normal vector, and a pressure is considered a negative stress.

This means that $\Psi_{ij} = -p\delta_{ij}$, where δ_{ij} is the *Kronecker delta,* which is 1 when the two indices are equal, 0 otherwise. In solids or in viscous fluids, the stress may also have a component in the plane, called a *shear stress.* In this case, Ψ will acquire off-diagonal components.

In an isotropic solid, the relation between stress and strain, or between Ψ and Φ, takes the form $\Psi_{ij} = [k\Phi_{kk} -(2\mu/3)\Phi_{kk}]\delta_{ij} + 2\mu\Phi_{ij}$. The quantity Φ_{kk} is the sum of the diagonal elements of Φ, also called its *trace.* It is equal to $\Delta V/V$, the relative volume expansion of the material, and called the *dilatation.*

The trace does not depend on the orientation of the axes, which should be clear from its meaning. In general, each of the six components of Ψ should depend on the six components of Φ, giving 36 constants of proportionality if we assume Hooke's Law, that stress is proportional to strain.

If the medium is isotropic, or the same in any direction, then these reduce to only two constants, the modulus of expansion k, $p = -k(\Delta V/V)$ (positive p causes a decrease in V), and the modulus of shear, μ. Young's modulus, for example, is $Y = 3k\mu/(k + \mu/3)$ and Poisson's ratio is $\sigma = (3k - 2\mu)/(6k + 2\mu)$. This is much simpler than it appears. Try contracting Ψ

and some normal vectors and see what stresses you get, for simple Φ's. In the case of a fluid, we find analogously that $\Psi_{ij} = [-p - (2\mu/3)\Phi_{kk}]\delta_{ij} + 2\mu\Phi_{ij}$, in terms of the pressure p and the absolute viscosity μ, which relates a shear stress to the rate of shear strain. The analogy is clear. $\Phi_{kk} = \mathrm{Tr}\ \Phi = \partial_k v_k = \mathrm{div}\ v$ is the dilatation.

For an incompressible fluid, this quantity is zero, and the stress tensor become even simpler. Unlike solids, fluids are almost always isotropic, so all this is less of an approximation. We are now able to express the stresses in a viscous fluid in terms of the fluid velocity.

EQUATIONS OF MOTION

The motion of a fluid in the Eulerian picture is defined by the velocity field v or v_i, which is the velocity of the particle of fluid that happens to be at P(x,y,z) at time t. A time dt later, this particle has moved off a distance v_i t and has been replaced by another. If we are interested in the time rate of change of some property of the particle of fluid, which we shall call F, then dF/dt must "follow the particle," or $dF/dt = \partial F/\partial t + v_i \partial_i F$.

This is called the *substantial derivative,* and is one of the first things explained in fluid mechanics. The mass of fluid must be conserved in the motion, or $d\rho/dt + \rho\partial_i v_i = 0$, where the dilatation appears again. To find how ρ as a point function behaves, we use the substantial derivative to find $\partial\rho/\partial t + v_i\partial_i\rho + \rho\partial_i v_i = 0$, or $\partial\rho/\partial t + \partial_i(\rho v_i)$. This is the *equation of continuity* that must hold for any possible motion.

Newton's Second Law for a particle of fluid is $\rho(dv_i/dt) = \rho F_i + \partial_j \Psi_{ij}$, where we have used the divergence theorem to get the force per unit volume from the stress tensor. ρF_i is an external force per unit mass, such as gravity.

The total force on the particle is the sum of the external force and the force given by the stress tensor. Using the expression of Ψ in terms of Φ, and the substantial derivative, we find: $\rho(\text{“}v_i/\partial t) + \rho v_k \partial_k v_i = \rho F_i - \text{“}_i p + (\mu/3)\partial_i(\text{“}_k v_k) + \mu\partial_k\partial_k v_i$.

This is called the Navier-Stokes equation, derived by Navier (1822), Poisson (1829), Saint-Venant (1843) and Stokes (1845) in successively more modern form. The second term on the left-hand side makes it nonlinear in v, which is extremely unpleasant if we are looking for analytical solutions.

If we specialize to the slow motion of an incompressible fluid, the Navier-Stokes equation becomes a bit simpler: $\rho(\partial v_i/\partial t) = \rho F_i - \partial_i p + \mu \partial_k$part;$_k v_i$.

VORTICITY

As elements of a solid body can rotate, so can elements of a fluid. If we antisymmetrize the rate-of-strain tensor, that is form $\Omega_{ij} = (\varepsilon_{ij} - \varepsilon_{ji})/2$, we eliminate the pure strain and are left with the rotation. This is an *antisymmetric tensor* with three independent (off-diagonal) components. These three components form a vector $(1/2)\omega = (1/2)$ curl v. curl v is the *vorticity* of the motion, a field quantity like any other.

From Stokes's Theorem, the line integral of the velocity around a closed curve is equal to the flux of vorticity through it. The diagram shows how the antisymmetric part of the strain is related to the angle of rotation.

Taking the time derivative gives the relation between vorticity and curl v. The lines of unit length and the strains in the diagram are actually multiplied by some small number so that the change in the lengths of the axes is infinitesimal, and the strains perpendicular to both. This is hard to represent clearly in a diagram.

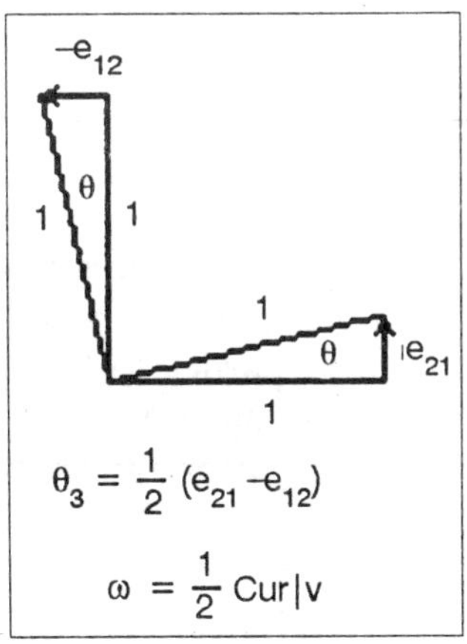

Fig: Rotation

If we take the curl of the Navier-Stokes equation (in its slow, incompressible form), we find that ∂(curl v)/∂t = μ div grad curl v, since curl commutes with div grad (write it out, if you don't believe it).

This is the same equation satisfied by the diffusion of heat for each component of the vorticity. We see that vorticity diffuses through the fluid with the diffusion coefficient equal to the viscosity. If the fluid is non-viscous, then ∂(curl v)/∂t = 0, and curl v = constant. This is a simple form of Kelvin's Vorticity Theorem, stating that for an ideal fluid, the vorticity is a constant. If the motion began from rest, then this says that the vorticity will be zero everywhere ever after, and the flow will be *irrotational*. Even the smallest amount of viscosity will allow vorticity to be created, and it may dominate the flow pattern. In the present case, with Poiseuille's and Stokes's Laws, we expect vorticity.

FLOW AROUND A SPHERE

The problem is shown at the right. A rigid sphere of radius a is placed in a uniform flow of velocity V in the x-direction at large distances. The problem is symmetrical about the x-axis, so the velocity in the xy-plane has no z component, and conditions are the same in any plane through the x-axis.

We assume the flow to be steady, slow and incompressible. The velocity must satisfy the equation of continuity, div v = 0, Newton's Second Law, div grad v = (1/μ) grad p, and must vanish on the surface of the sphere. At great distances, its x-component must be V.

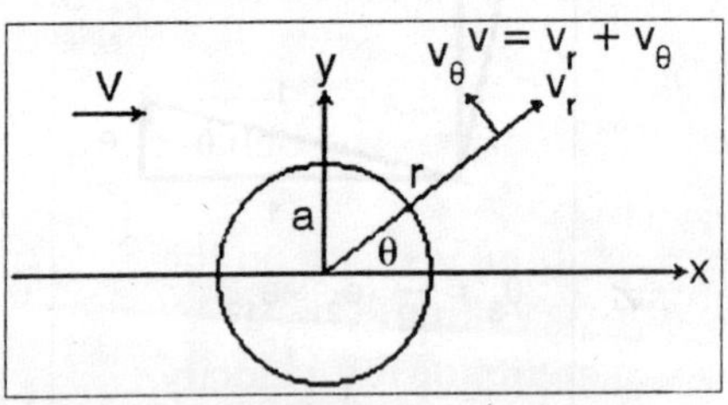

Fig: Sphere in a Uniform Flow

We'll use the vorticity, and a little luck, to solve this

problem. Taking the curl of the equation of motion, we get div grad (curl v) = 0, so curl v is a solution of Laplace's equation.

From the axial symmetry of the problem, we can put curl v = f(r,θ)(-jsin φ + kcos φ), where j and k are unit vectors in the y and z (out of the page) directions, and φ is the azimuthal angle around the x-axis. Then, appropriate solutions of Laplace's equation are r sin θ and sin θ/r^2, where the choice of sin θ instead of cos θ has been made so that the solution will vanish on the axis.

Substitution in Laplace's equation will verify that these are, indeed, solutions. The first solution must be thrown out because it is infinite at large distances. Then, curl v = $A(-jz/r^3 + ky/r^3)$ = curl (iA/r), where z = r sin θ sin φ, y = r sin θ cos φ, and i is a unit vector in the x-direction. Therefore, we must have v = i(2A/r) + grad ψ, where ψ is some scalar function. The equation of continuity then gives div v = -2A cos θ/r + div grad ψ = 0, or div grad (A cos θ + ψ) = 0.

This will be satisfied provided ψ = -A cos θ + φ, where div grad φ = 0. Since φ satisfies Laplace's equation, possible solutions are B r cos θ and C cos θ/r^2, where solutions proportional to cos θ have been selected to agree with the first term in angular dependence.

Finally, then, we have v = i(2A/r) – A grad cos θ + B grad (r cos θ) + C grad (cos θ/r^2). When we evaluate the gradients, we have an expression for the velocity, in terms of its radial and angular components.

This expression for v gives the required vorticity, satisfies the equation of continuity, and now if it can be made to satisfy the boundary condtions by proper choices of A, B and C, it will be the desired solution. The behaviour at large distances requires B = V.

The vanishing of the velocity on the surface of the sphere demands that A = -3aV/4 and C = $-a^3$ V/4.

For the components of the velocity, we have $v_r = V(1 - 3a/2r + a^3/2r^3)\cos\theta$ and $v_\theta = -V(1 - 3a/4r - a^3/4r^3)\sin\theta$. The z-component of curl v in the xy-plane is $-(3aV/4r^2)\sin\theta$.

Now we can use the stress tensor to find the stresses on

the surface of the sphere, and by integrating them over the sphere to find the drag.

This is made much easier by choosing special x,y and z axes at a point on the sphere at angle θ with the x-axis, with the x-axis radially outward, the y axis in the direction of increasing θ, and the z-axis normal to the xy-plane, as shown in the diagram.

Then, $\partial v_x/\partial x = \partial v_r/r\partial\theta$, $\partial v_y/\partial y = \partial v_r/r\partial\theta + v_r/r$, $\partial v_y/\partial x = \partial v_\theta/\partial r$, $\partial v_y/\partial y = \partial v_\theta/r\partial\theta - v_\theta/r$. Don't forget that in the second and fourth equation the unit vector also changes, and gives a contribution.

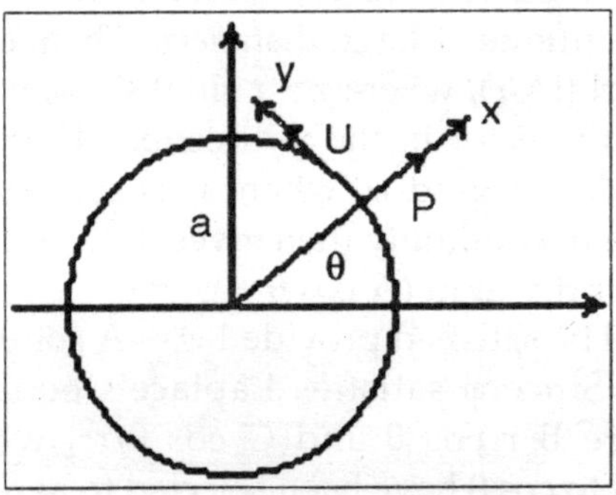

Fig: Temporary Axes

The only contributions that do not vanish on the surface of the sphere are the normal pressure, $P = (3\mu V/2a)\cos\theta$, and the shear stress $U = \mu(\partial v_y/\partial x) = -(3\mu V/2a)\sin\theta$.

The net stress on unit area is $P\cos\theta - U\sin\theta = 3\mu V/2a$ in the x-direction, so that the total drag becomes $4\pi a^2(3\mu V/2a) = 6\pi\mu aV$, the desired result. P and U can be separately integrated over the area, of course, and it is found that the pressure contributes 1/3, the shear 2/3 of the total drag.

Note that there is a positive pressure over the face of the sphere facing the flow, and a negative pressure over the other face. The shear force is a maximum at the equator, and is zero on the axis. Page, and other authors, copy Lamb's pictures of the stream functions for the irrotational flow around a sphere and for the viscous flow, and remark on how very different they are, repeating Lamb.

Indeed they are, but they refer to two different situations, the first where the sphere is moving, and the second where the fluid is moving. The actual difference is not as great as may appear from these diagrams, but, of course, there is a significant difference nonetheless.

FLOW THROUGH A TUBE

Poiseuille's problem is quite a bit easier. We can solve it using the definition of viscosity and a little statics. Consider a cylinder of fluid of thickness dr at a radius r, and of length L, and let this cylinder move at velocity v = v(r).

The fluid on the side towards the wall moves a little slower, that towards the centre a little faster.

The velocity gradient causes the outer fluid to retard our cylinder, the inner fluid to urge it forward. In equilibrium, the velocity gradient will arrange itself so that the net force is balanced by the pressure difference at the ends of the cylinder.

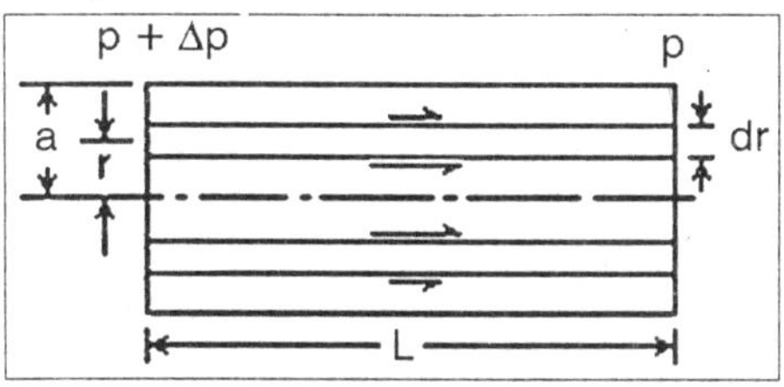

Fig: Poiseuille Flow

This is expressed mathematically by $(d/dr)[2\pi rL\mu(dv/dr)]dr = -2\pi r\Delta p\ dr$. Integrating once, we find $2\pi rL\mu(dv/dr) = -2\pi(r^2/2)\Delta p$, or $dv/dr = -(\Delta p/2L\mu)\ r + A$, where A is a constant. By symmetry, when r = 0, dv/dr = 0, so A = 0.

Integrating a second time, $v(r) = -(\Delta p/4L\mu)r^2 + B = 0$, where B is a second constant of integration. Since v = 0 when r = a, the evaluation of B gives $v(r) = (\Delta p/4L\mu)(a^2 - r^2)$, which is the parabolic velocity profile of Poiseuille flow. As with the sphere, the vorticity forms rings around the x-axis. It is easy to find the vorticity from the velocity profile.

The rate of discharge Q is found by integrating the velocity over the cross section, $Q = 2\pi\int v(r)rdr$. The integral is easily performed after making the substitution $u = r^2$, with the result that $Q = (\pi a^4/8L\mu)\Delta p$. The rate of discharge is proportional to the pressure difference Δp divided by the length L, or the pressure gradient, to the fourth power of the radius of the tube, and inversely proportional to the absolute viscosity. The mass discharge is proportional to the kinematic viscosity μ/ρ.

FLOW IN A PIPE

Flow in cylindrical pipes is a subject of great importance in engineering, and a great deal of empirical data has been built up that allows the solution of most practical problems with relative ease.

In general, practical flow in pipes is turbulent, not laminar. The engineer speaks of "head loss" instead of pressure loss. Pressure and head are directly related through $p = \rho gh = \gamma h$, where p is pressure in dyne/cm^2, h is head in cm, ρ is the density in gm/cm^3 and γ is specific weight in dynes/cm^3. Of course, engineers use other units, such as lb/ft^2, ft, slug/ft^3 and lb/ft^3. The head can be measured directly by a manometre at any point, though actual manometres are seldom used these days. A certain volume rate of flow Q through a length L of pipe is associated with a loss in head of h/L, a dimensionless ratio.

The engineer uses the equation $h = f(L/D)(V^2/2g)$, where D is the inside diametre of the pipe, V is the average velocity Q/A, where $A = \pi D^2/4$, h the head loss in a length L of pipe, and f the *Fanning friction factor*. The friction factor is a function of the Reynolds number $DV\mu/\rho$ and, for turbulent flow, of the relative roughness of the pipe, e/D, where e is the absolute roughness. The usual connection is given in the famous chart by L. F. Moody, which can be found in Daugherty. Turbulent flow is usual when the Reynolds number exceeds 2000. This is much larger than for a sphere, because the pipe tends to keep things smooth. Below 2000, Poiseuille flow governs.

Poiseuille's equation, expressed in terms of head loss, is $h = 32\mu LV/\gamma D^2$. Comparing this with the Fanning equation, we find that $f = 64/R$, where R is the Reynolds number. This is represented by a line in the Moody chart showing that f decreases steadily to about 0.032, where the critical Reynolds number is reached, and then jumps abruptly to perhaps twice as much before it begins to decrease again as R increases.

TENSORS

An ellipsoid is a quadric surface described in canonical form by the equation $ax + by^2 + cz^2 = 1$. The lengths of the semiaxes of the ellipsoid are $1/a^{1/2}$, $1/b^{1/2}$ and $1/c^{1/2}$ along the x, y and z axes, respectively. If b = c, the ellipsoid degenerates into a spheroid, and if a = b = c, into a sphere. The centre of the ellipsoid is at the origin.

If the axes are rotated to a general position, then the equation for the ellipsoid can be written $ax^2 + by^2 + cz^2 + dyz + ezx + fxy = 1$, where a,b and c are new constants, as well as d, e, f. The absence of linear terms means that we keep the origin at the centre of the ellipse, since displacing the centre of the ellipsoid will not add anything essential. There are six constants in this expression, and they specify the shape and size of the ellipse and the directions of the axes, which remain orthogonal. Any section of an ellipsoid by a plane is an ellipse, notably the principal sections that contain its axes.

A rank-2 symmetrical tensor Φ_{ij} has six components, three diagonal and three off-diagonal, which occur in symmetric pairs. The equation $x_i\Phi_{ij}x_j = 1$ is the equation of an ellipsoid associated with the tensor. A radius vector x is expressed as its components x_i, i = 1, 2, 3 instead of x, y, z.

It is possible to find the rotation that transforms a symmetric rank-2 tensor into the *diagonal* form $\Phi_{ij} = a^{(i)}\delta_{ij}$. This will also transform the ellipsoid into canonical form and give the lengths of the semiaxes. To do this, let n_i be a unit vector (the components are direction cosines) and form the equation $\Phi_{ij}n_j = \lambda n_j$, where λ is some constant, which we may have to determine. This expresses the condition that the vector resulting when the tensor Φ "operates" on n_i gives a vector in the same direction, but of possibly different magnitude.

It is the fundamental duty of rank-2 tensors to operate in this way on one vector (by this is meant multiplication followed by contraction) to give another vector. The expression is really three simultaneous equations for the three direction cosines n_i. Therefore, we have the three equations $(\Phi_{ij} - \lambda\delta_{ij})n_j = 0$ for the three values of i. These are *homogeneous* equations, and only have the solutions $n_i = 0$, unless the determinant of the coefficients vanishes. If it does, we can solve for the ratios of two of the direction cosines to the remaining one, and then determine them all by the condition that the sum of their squares must be unity.

The constant λ must be chosen so that the determinant of the coefficients vanishes, $|\Phi_{ij} - \lambda\delta_{ij}| = 0$. Since this equation is of third order, there are three roots for λ. If the three roots are equal, then the tensor is already diagonal, and is the constant tensor $\lambda\delta_{ij}$. The corresponding ellipsoid is just a sphere. If two roots are equal, then the remaining root gives the direction of an axis of symmetry, and the orientation of the axes perpendicular to this axis can be chosen any way one likes.

The ellipsoid is then a spheroid, and,

$$\Phi_{11} = \lambda(1),$$

$$\Phi_{22} = \Phi_{33} = \lambda(2) = \lambda(3).$$

Finally, in the general case all three roots are unequal, and we have an ellipsoid with unequal axes, and the diagonal elements of Φ are the three values of λ. If the three values of λ are unequal, the resulting *principal axes* will all be *orthogonal*, that is, at right angles to each other. If two values are equal, then the axis will be orthogonal to the plane of the other two axes. The values of λ are called *eigenvalues* ("own values") and the unit vectors belonging to them are called *eigenvectors*. Working with principal axes simplifies the algebra greatly, and we usually assume that any symmetrical rank-2 tensor has been *diagonalized* in the way we have discussed. All of these matters are discussed at length in texts on linear algebra.

We now have enough theoretical background to consider the most important rank-2 tensors met with in classical physics. We'll discuss the inertia tensor, the dielectric tensor, the strain tensor, and the stress tensor in this article.

Not in detail, of course, but enough to show how the tensors are defined and used, making their application easier.

There are good numerical methods for the diagonalization of matrices, and the basic theorems are proved in any quantum mechanics text.

THE INERTIA TENSOR AND RIGID BODY MOTION

The inertia tensor in the dynamics of rigid bodies is an excellent example of a rank-2 tensor where the associated ellipsoid aids in the visualization of the motion. We shall start from first principles, using index notation, and find the motion of a free rigid body—that is, a body under the action of no external forces.

Any rigid body can be considered as an assembly of mass elements m with fixed positions x_i relative to an orgin O, in a system of coordinates fixed in the body. The origin O can be a fixed point in space and the body, or the centre of mass of the body. In either case, the motion separates into translational and rotational parts. We are not concerned with the translational parts here, and will consider only the rotation.

The most general motion of the body is then a rotation with angular velocity ω_i. The direction of the vector specifies the direction of the axis of rotation. The angular velocity may change in magnitude and direction with time, in a quite general manner. The body coordinates rotate in the same way, and so do not form an inertial system.

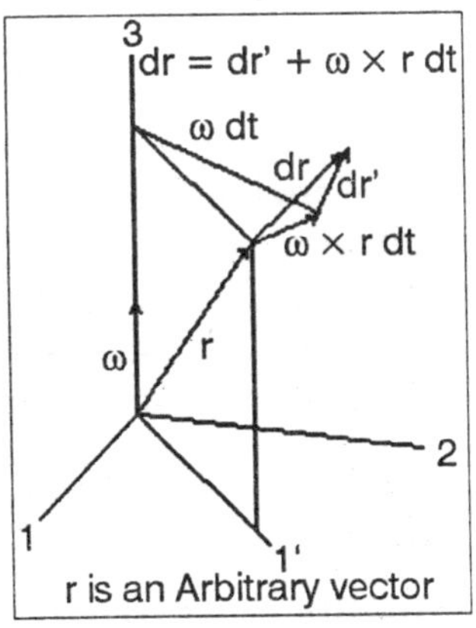

Fig: Increments in Fixed and Rotating Systems

The rate of change of any vector quantity v_i in a fixed coordinate system with origin O is $[dv_i/dt]_{fix} = [dv_i/dt]_{rot} + \varepsilon_{ijk}x_jv_k$. This is merely the relation that in vector notation is,

$$[dv/dt]_{fix} = [dv/dt]_{rot} + r \times v,$$

which is illustrated in the diagram, which applies to any vector, not just a radius vector. Accordingly, the velocity of an element of mass m in the inertial system is $v_i = \varepsilon_{ijk}\omega_jx_k$, since it is fixed in the rotating system.

Summing over all mass elements, the angular momentum $H_i = \Sigma m\ \varepsilon_{ijk}x_j\ \varepsilon_{klm}\omega_lx_m$. Using the value of the contraction of two antisymmetric tensor densities, we find that $H_i = [\Sigma m(x_kx_k\delta_{ij} - x_ix_j]\ \omega_j$, The rank-2 symmetric tensor multiplying ω_j is the *inertia tensor* I_{ij} of the body.

The diagonal elements are called *moments of inertia,* and the off-diagonal elements *products of inertia.* We note that the angular momentum is not necessarily in the direction of the angular velocity. Now we find the kinetic energy T of the body by adding up the kinetic energies of each element of mass: $2T = \Sigma v_kv_k$. Using the expression for the space velocity, we again get a contraction of two antisymmetric tensor densities. In fact, $2T = \Sigma m\ \varepsilon_{ijk}\omega_jr_k\ \varepsilon_{irs}\omega_rr_s = I_{jr}\omega_j\omega_r$. This work is a good example of how much easier it is to use the index notation than vector notation in these things.

Newton's Second Law can be applied in the fixed system to each of the elements of mass, and we find that the net torque $L_i = \Sigma\varepsilon_{ijk}x_jF_k$ is the rate of change of angular momentum H_i in the fixed system. From the above equations, we have $L_i = I_{ij}(d\omega_j/dt) + I_{kl}\varepsilon_{ijk}\omega_j\omega_k$. These are three differential equations for the components of the angular velocity of the body relative to the body axes. It is always possible to diagonalize the inertia tensor by a suitable choice of the orientation of the body axes. Let us assume this has been done, and the principal moments of inertia are $A^{(i)}$. The index is in parentheses because it is not a tensor index, merely an identifying label.

For ease of writing, we may also represent these quantities by A, B and C. Then, $I_{ij} = A^{(i)}\delta_{ij}$, and the equations of motion of the preceding paragraph become $L_i = A^{(i)}(d\omega_i/dt) + \varepsilon_{ijk}A^{(k)}\omega_j\omega_k$. These equations are known as *Euler's equations.*

Now we can consider the problem of a body moving freely with one point fixed, which can be taken as the centre of mass. An asteroid moving in space is a good example of such a body, since the torques exerted by unequal solar attraction are small and have little effect if the body is rotating rapidly.

Such torques indeed cause the axis of the earth to precess, but this is a slow effect. We can safely neglect such torques with an asteroid rotating with periods of hours or less.

First of all, the kinetic energy T of the body must be constant, so $2T = H_i\omega_i = I_{ij}\omega_i\omega_j$ = constant.

The angular momentum is also constant, since no external torques act, so this says that the component of the angular velocity in the direction of H is also a constant. If we consider a space in which the components of the angular velocity are the coordinates, then this equation also defines an ellipsoid in this space, which is similar to the inertia ellipsoid. We shall call this the *energy ellipsoid*, since its size depends on the kinetic energy of the motion. We shall choose the body axes so that the inertia tensor is diagonalized, so the body axes are also the symmetry axes of the energy tensor. The semiaxes of the energy ellipsoid are $\Phi(2T/A^{(k)})$. The angular velocity is a vector from the origin O to a point on the energy ellipsoid P, as shown in the diagram at the left.

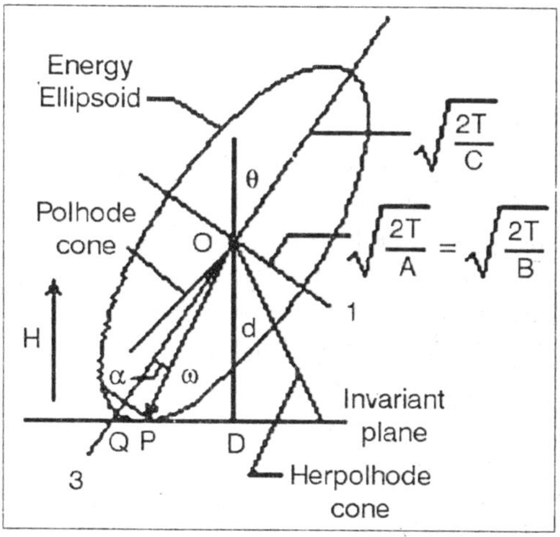

Fig: Free Motion of a Rigid Body

If the equation for the energy ellipsoid is differentiated, we find that $I_{ij}\omega_j d\omega_i = 0$. Now $d\omega_i$ is a vector in the tangent plane at P, so this says that the angular momentum is perpendicular to the tangent plane. Because the angular momentum is constant, this means that the orientation of the tangent plane remains constant during the motion. Also, since the component of the angular velocity in the direction of the angular momentum is constant, the distance d from the origin O to the tangent plane is constant.

Therefore, we find the important result that the tangent plane to the energy ellipsoid is fixed in space. We call it the *invariant plane*. Since the angular velocity passes through the point of contact P, there is no relative motion between the energy ellipsoid and the invariant plane: the energy ellipsoid must roll on the invariant plane without slipping.

The angular velocity must terminate on the energy ellipsoid in any case. For a certain d, it must also lie on the surface obtained by squaring the relation Hd = 2T, which is $\Sigma(A^{(i)}d\omega_i)^2 = 4T^2$. Eliminating the constant terms between these two equations, we find that $\Sigma(A^{(i)}/2T)[(1/d^2) - (A^{(i)}/2T)]\omega_i^2 = 0$, which is the equation of a cone (not in general a circular one) with centre at O, on which the angular velocity must lie. This cone intersects the energy ellipsoid in a curve called the *polhode* (Greek: "path of the pole").

This cone rolls without slipping on a similar cone fixed in space determined by the path of the angular velocity in the invariant plane, which is called the *herpolhode*. We have now arrived at a complete description of the motion of the body, making use of the inertia ellipsoid.

Let's now specialize to the case frequently met when two of the moments of inertia are the same, say A = B, with C different. The case when C < A is shown in the diagram, where the inertia ellipsoid is prolate, and the 3-axis is the symmetry axis. Now the polhode and herpolhode determine circular cones that roll on each other without slipping, and the motion can be easily visualized. The component of angular velocity Ω along the symmetry axis is called the *spin*. It is given by $\Omega = \omega \cos \alpha$.

The radius of the polhode is ω sinα, and the radius of the herpolhode is ω sin $(\theta - \alpha)$. The angle between the symmetry axis and the direction of H is θ. It is related to α by tanα = (C/A) tanθ. The rotation of the symmetry axis around the direction of H is called *precession*, and its amount is $d\psi/dt = C\Omega/(A \cos\theta)$.

In finding the inertia tensor for bodies made up of simple geometric shapes, the *Parallel-Axis Theorem* is useful.

It says that the moment of inertia about an axis parallel to an axis through the centre of gravity is equal to the moment about the axis through the centre of gravity plus the total mass of the body times the square of the distance between the axes, or $I_{ii} = I_{ii}' + md^2$. For products of inertia (off-diagonal elements), the product of inertia is the product of inertia referred to the centre of gravity plus the mass times the product of the coordinate differences.

That is, $I_{ij} = I_{ij}' + mxy$. Tables of the moments of inertia of simple figures can be found in handbooks and in texts of engineering mechanics. The moment of inertia of a parallelepiped about an axis through the centroid perpendicular to a face is $(m/12)(a^2 + b^2)$, where a and b are the sides of the face. The centroidal moment of inertia of a sphere is $(2/5)mr^2$, and of a disk relative to an axis perpendicular to its plane, like a wheel and axle, $(1/2)mr^2$.

THE STRAIN AND STRESS TENSORS AND ELASTIC CONSTANTS

Consider a solid elastic medium that is deformed by forces applied to its boundaries, or by forces exerted directly on the material by external influences. In response to these forces, a general point originally at x_i before the forces are applied moves to a point $x_i + \hat{\imath}_i$.

The displacement $\hat{\imath}_i$ is not necessarily small, but its derivatives E_{ij} = "$_i\hat{\imath}_j$ we presume are much less than unity. These quantities are called *strains*, and form a rank-2 tensor that is a function of position. It is convenient to separate E_{ij} into symmetric and antisymmetric parts, $E_{ij} = e_{ij} + \Omega_{ij}$. The symmetric part e_{ij} is called the *pure strain* tensor, while the antisymmetric part Ω_{ij} represents the rotation due to the deformation. The vector associated with Ω_{ij} gives the rotational axis and the angle of rotation. We will be mainly interested in the pure strain, which represents the deformation of the medium.

If the medium returns to its initial state when the forces are removed, the deformation is called *elastic,* and all the strain energy put into the body is recovered. If the deformation is too large, the body does not return to the initial state, mechanical energy is dissipated, and the deformation is called *plastic*.

For solid bodies, deformation is usually elastic for sufficiently small strains. We shall deal with this case exclusively here. Since the pure strain is symmetric, it can be diagonalized. Let the diagonal elements be a, b, and c.

Then, a small parallelepiped with sides dx, dy and dz and volume V = dxdydz becomes a small parallelepiped with sides (1 + a)dx, (1 + b)dy and (1 + c)dz after deformation, of volume V + dV = (1 + a + b + c)dxdydz, so that dV = (a + b + c)V.

Quantities of the order of squares of strains have been neglected. The sum a + b + c = dV/V is the relative increase in volume at the point, called the *dilatation*. Now, a + b + c is just the trace of the strain tensor, e_{ii} = "$\hat{\imath}_{i}$, which we know is invariant under rotation. As an example, consider a body strained as shown in the figure by a shearing force.

The only nonzero component of the strain tensor is E_{12} = a, where a is a small quantity equal to the angle of strain θ. Then the pure strain tensor has $e_{12} = e_{21}$ = a, and the rotation tensor has $\Omega_{12} = -\Omega_{21}$ = a.

We note that the deformation is indeed equal to a rotation of a radians about the 3-axis plus an extension along the 2-axis and a contraction along the 1-axis. Indeed, the pure strain tensor has eigenvalues ±a/2. The trace is zero, so there is no dilatation. This sort of deformation is called *pure shear*.

Now we must consider the causes of deformation and rotation, stresses. If we imagine a plane surface in the body and that the material on one side of the plane surface is removed, the forces that it exerted on the material on the other side can be represented by forces acting on this plane surface.

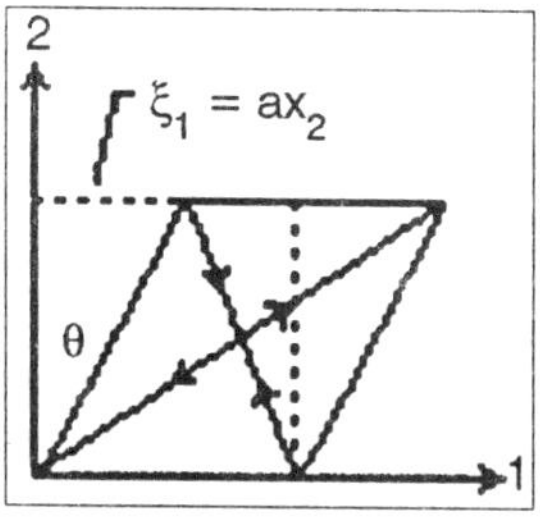

Fig. Deformation in Shear

Let the unit normal vector to the surface be taken as the normal pointing towards the removed material, or outwards. The normal force is positive when it is a pull, or in the direction of the normal, or a *tension*. There may also be a force in the plane of the surface, perpendicular to the normal, called a *shear force*, and in any particular case the positive directions of its two components in the surface will be agreed upon.

A force per unit area of surface is a *stress*. If a hydrostatic pressure p acts on the surface, then the normal stress is -p and the shear stresses are zero. The stresses will, in general, depend on the direction of the normal. For any normal vector n_i, we will have stresses F_i that are a function of the n_i. This defines a tensor S_{ij} such that $F_i = S_{ij}n_j$, called the *stress tensor*. That this is a satisfactory definition is can be seen by considering the equilibrium of a small tetrahedron, as shown in the figure at the left. The three components of the force on the inclined face of area dS are balanced by equal and opposite forces on the other three sides. Note that the area of each side is n_idS, and the force on it will be in the opposite direction to the force on the face dS. To prove this, write out the forces in the 1, 2 and 3 directions separately and show that they balance.

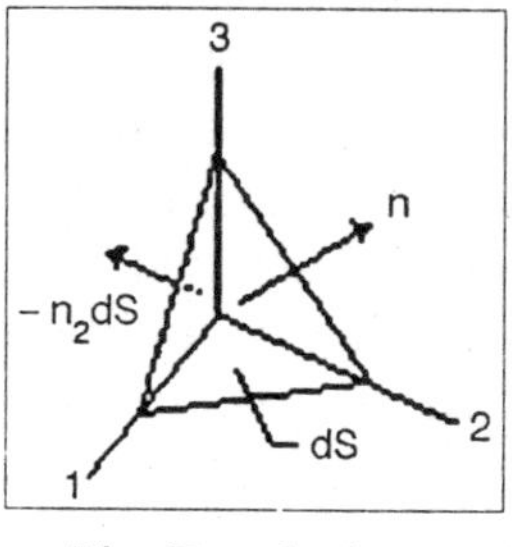

Fig. Tetrahedron

The stress tensor must be symmetric, or there would be unbalanced torques on small areas that would cause them to rotate. The absence of such rotation causes the stress tensor to be symmetric.

In the figure, we take the moments of the forces tending to rotate the cube about the 3-axis that come from the 1-2 component of the tensor, showing that $S_{12} = S_{21}$. The same holds for each of the other off-diagonal elements.

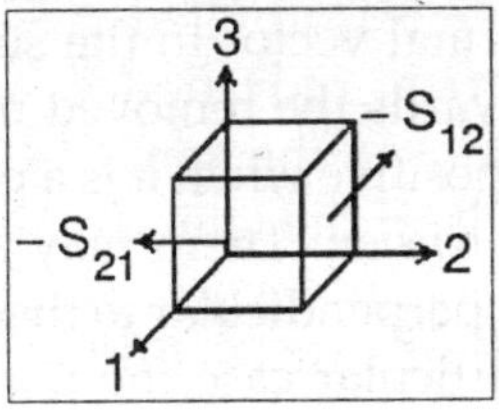

Fig: Symmetry of the Stress Tensor

We now look for the connection between the stress and pure strain tensors. Since the strains are small compared to unity, we assume that Hooke's Law holds, and the stresses are linear functions of the strains. Actually, Hooke's Law is always valid for *elastic* deformations, since the strain is still small when inelasticity begins.

A systematic way to do this is to consider a deformation energy F defined by "F/"$e_{ij} = S_{ij}$.

Hooke's Law then requires that S be proportional to e, or that F be a homogeneous quadratic function of the strains. The most general isotropic quadratic homogeneous function of the strains is $F = (1/2)\lambda(e_{kk})^2 + \mu(e_{ij})^2$, where the constants λ and μ are called the *Lamé* coefficients. If we differentiate F, we find $S_{ij} = \lambda e_{kk}\delta_{ij} + 2\mu e_{ij}$, which is the relation between stress and strain for an isotropic, elastic body.

Poisson took $\lambda = \mu$, as did Cauchy and others who thought there should be only one elastic constant for an isotropic medium. Their theories were based on central forces between supposed "molecules" that have little in common with our modern molecules. There was an extended debate between those who demanded more elastic constants and those who demanded fewer.

In the general case (as we shall discuss below) 21 constants were required by those who favoured more, and only 18 by those favouring less. Time and experiment have come down on the side of those who think there must be the larger number.

Elastic constants useful in specific problems can be defined in terms of λ and μ. If we assume an isotropic pressure p, then $-p = \lambda e_{kk} + 2\mu e_{(k)(k)} = (\lambda + 2\mu/3)dV/V$.

The *bulk modulus* k is defined by dV/V = -p/k, so k = λ + 2μ/3. If an axial force F is applied to a rod of length L and cross-sectional area A, the rod lengthens by ΔL = FL/AY, where Y is *Young's modulus*. At the same time, the rod contracts laterally, and the ratio of the lateral strain to the longitudinal strain is σ, *Poisson's Ratio*.

Under these conditions, we find that σ = λ/(2λ + 2μ) and Y = 2(σ + 1)μ. Poisson's ratio can be between 0 and 1/2, but is usually not far from 1/4, which Poisson assumed was the case. It is 1/2 for a body with no shear strength, such as a liquid.

The modulus μ itself is the ratio of a shearing stress to the shearing strain (an angle), so is called the *modulus of rigidity*. A quantity called a *modulus* has the dimensions of a stress, it should be noted.

Steel has Y = 30 × 10^6 psi or 2.07 × 10^{12} dyne/cm^2, μ = 12 × 10^6 psi or 8.27 × 10^{11} dyne/cm^2, and σ = 0.27. Its bulk modulus, calculated from these figures, is k = 20 × 10^6 psi or 1.38 × 10^{12} dyne/cm^2. Water, on the other hand, has μ = 0, since it lacks rigidity, and k = 3.16 × 10^5 psi or 2.18 × 10^{10} dyne/cm^2.

It is about 63 times more compressible than steel. As for gases, dV/V = -dp/p for an ideal gas, so k = p. At one atmosphere, this is 1.47 psi or 1.013 × 10^6 dyne/cm^2, about 22,000 times more compressible than water.

Rubber is a curious material with σ H" 0.5, which implies that it has very low rigidity, a kind of "solid fluid." For rubber, Y = 100 to 600 psi, and μ = 30 to 200 psi, so its rigidity is indeed small. Glass is on the opposite pole, with Y = 10 × 10^6 psi and μ = 5 × 10^6 psi. Its Poisson's ratio is low, 0.20 to 0.27, which implies that it is quite rigid compared to its longitudinal elasticity.

A medium whose properties depend on direction is often

called *anisotropic,* but a less barbarous word is *aelotropic* (Greek: change-turning). The above expression for F suggests the generalization to an aelotropic medium by setting $F = (1/2)\lambda_{ijkl}e_{ij}e_{kl}$, which gives $S_{1j} = \lambda_{ijkl}e_{kl}$. The rank-4 tensor λ_{ijkl} is called the *elastic constant tensor,* an operator connecting two rank-2 tensors just as a rank-2 tensor is an operator connecting two vectors.

In general, it has 81 components, but there is a great deal of symmetry here, since the energy is unchanged if i,j or k,l are interchanged (the strain tensors are symmetric), or if i,k and j,l are simultaneously interchanged.

As a result, there are only 21 independent components at worst. It is confusing to work this out, but all 81 components can be listed, and equal ones sorted out, as a last resort.

The symmetry of crystals will introduce more relations between the constants, reducing their number. Even in a triclinic crystal, the least symmetric, the possibility of choosing convenient coordinates reduces the number to 18.

A monoclinic crystal has 12, an orthorhombic crystal 9. A tetragonal crystal has 6, as does a trigonal (rhombohedral) crystal like calcite or quartz.

A hexagonal crystal has 5, and a cubic crystal 3. Cubic crystals are often considered to be isotropic in many respects, but they have an extra elastic constant that truly isotropic materials do not have. The constant corresponding to λ splits into two, one the coefficient of the squares of the diagonal elements of e, the other the coefficient of the cross products.

This reduction in the number of constants is found by applying the symmetry elements of the crystals, which turns some constants into their negatives, so they must vanish. This is an interesting point, but we cannot go into it further at this time.

An interesting application of this theory is to the propagation of mechanical waves in solids, which has application to seismology, nondestructive testing, and other fields. Such waves are almost always elastic, and have an interesting variety of properties that have been intensively studied.

The theory is generally more complicated than that of electromagnetic waves, and includes the subject of surface waves (Rayleigh and Love waves), as well as waves in an infinite medium.

THE DIELECTRIC TENSOR AND CRYSTAL OPTICS

The relation between the electric displacement D (statcoulombs/cm$_2$) and the electric intensity E (statvolts/cm), considered in the first approximation as a proportionality, will in general be given by $D_i = \varepsilon_{ij}E_j$, where the dielectric tensor satisfies $\varepsilon_{ij} = \varepsilon_{ji}$.

The requirement of symmetry comes from several sources, one of which is simply that the tensor should be diagonalizable by an ordinary rotation, which establishes the three orthogonal principal axes of polarization, and the three principal dielectric constants, its eigenvalues. This yields $D_{(k)} = \varepsilon_{(k)}E_{(k)}$, where k = 1, 2 and 3. An electric intensity in a principal direction gives a displacement in the same direction. If all three directions are equivalent in the medium, then the three constants are equal: $\varepsilon_{(k)} = \varepsilon$, and an electric intensity in any direction creates a displacement in the same direction; the medium is then *isotropic*.

In crystals belonging to the cubic or isometric system, three equivalent axes can be found, so the medium is isotropic. In the trigonal (rhombohedral), tetragonal and hexagonal systems, two equivalent axes perpendicular to the third can be found. If the unique axis is the 3-axis, then $\varepsilon_{(1)} = \varepsilon_2$, while ε_3 is different. In the orthorhombic, monoclinic and triclinic systems, all three principal dielectric constants may differ. In any case, dielectric constants may *accidentally* be closely the same, so the medium will resemble a more symmetric one.

Noncrystalline materials, such as polymer sheets, may be anisotropic as well. Nonisotropic transparent materials may show *double refraction,* discovered in 1669 by Bartholinus in Iceland spar, or transparent calcite, $CaCO_3$, and first explained by Huygens not long afterwards. Such materials are also called *birefringent*. We must first discuss the propagation of plane electromagnetic waves in an anisotropic medium. We shall assume throughout that B = H, or $\mu = 1$, which is true in transparent crystals. Assume that the wave depends on space and time through a factor $\exp[i\omega(t - nr\cdot s/c)]$, where ω is the angular frequency, $v = c/n$ is the phase velocity, r the position vector and s is the unit vector normal to the wavefront. Then we can replace time derivatives by $i\omega$ and del by $-i\omega ns/c$ in Maxwell's equations. We find: $s\cdot D = 0$, $s\cdot H = 0$, $ns \times H = -D$, and $ns \times E = H$. All the terms in these expressions are constant vector amplitudes.

The spatial relations between the vectors used to describe the wave are shown in the diagram at the right. The wavefront is the plane of H and D, with s the wavefront normal. E lies in the plane of s and D making an angle α with D. Since browsers do not yet support the perpendicular symbol, the electric intensity along D will be represented in the text by E+. The Poynting vector S is in the direction of t, the ray unit vector. This gives the direction of energy flow, which is not normal to the wavefront. The phase velocity is vs, with $v = c/n$, and the ray velocity is $v_r t$, with $v_r = c/n_r$. The electric energy density $w_e = E\cdot D/8\pi = (n/8\pi)H\cdot(s \times E)$ and the magnetic energy density $w_m = H\cdot B/8\pi$ is given by exactly the same expression, so the total field energy $w = (n/c)s\cdot S$, where $S = (c/4\pi)(E \times H)$ is the Poynting vector. Now, $v_r = S/w$, so $v = v_r s\cdot t = v_r \cos\alpha$. The indices of refraction are then related by $n_r = n \cos\alpha$.

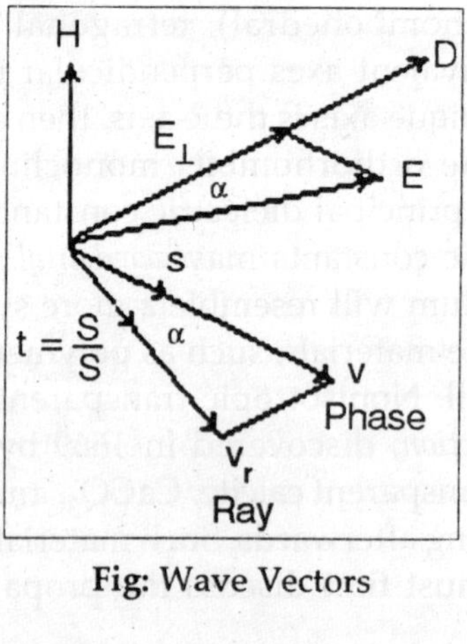

Fig: Wave Vectors

The magnetic field can be eliminated between the Maxwell curl equations with the important result that $D = n^2[E - s(s \cdot E)] = n^2E+$. This is a relation between E and D that must be satisfied by the wave, and is completely independent of the relation between them given by the dielectric tensor.

Both relations must be satisfied in the wave. Since $E+ = (E \cdot D)D/d^2$, this relation can also be written as $n^2 = D^2/(E \cdot D)$. Similarly, it can be shown that $n_r^2 = (D \cdot E)/E^2$.

Applying the diagonalized dielectric tensor (which means everything must be referred to principal axes) we find $\varepsilon_k E_k = n^2[E_k - s_k(s \cdot E)]$.

The k here is not a tensor index, merely an indentifier of one of the three components. Written out fully, we find three homogeneous equations for the field components E_k. Such a system can have non-zero solutions only if the determinant of the coefficients vanishes.

This condition will determine the value of n, or the phase velocity, for which both Maxwell's equations and the dielectric relation are satisfied. Fresnel discovered a cunning way to do this that yields a symmetrical equation for v in terms of the components of the wave normal vector.

Rearrange the equation to $E_k = [n^2 s_k/(n^2{}_k)]$ $(s \cdot E)$. This is valid if the quantity in parentheses does not vanish. Now multiply by s_k and add the equations for k = 1, 2 and 3. The dot product cancels, and we find a sum of three similar terms that adds to 1.

Now the sum of the squares of s_k is also 1, so subtract this from both sides and combine the terms with the same s_k's. Now change from n's and ε's to the velocities. The final result is $s_1^2/(v^2 - v_1^2) + s_2^2/(v^2 - v_2^2) + s_3^2/(v^2 - v_3^2)$.

This is *Fresnel's equation of wave normals.* To use it, multiply by the denominators to clear of fractions. The result is a quadratic equation in v^2, which gives two values for any value of s. The squares mean that the wave can travel in either direction. We will see that each pair of solutions for the same s gives us two plane-polarized waves polarized at right angles to each other which, in general, travel at different velocities.

A similar process in terms of t instead of s yields *Fresnel's ray equation,* which gives us the ray velocity v_r in terms of the ray unit vector. This can be found in Born and Wolf.

What is more interesting is to find t in terms of s and the phase velocity. The result is $t_k = s_k\{[v^2 + g^2/(v^2 - v_k^2)]/\Phi(v^4 + g^2)\}$, where $g^2 = v^2(v_r^2 - v^2) = \{[s_1/(v_1^2 - v^2)]^2 + [s_2/(v_2^2 - v^2)]^2 + [s_3/(v_3^2 - v^2)]^2\}^{-1}$. From these formulas, we can find the ray vector and the ray (energy) velocity. When the phase and ray velocities re known, then the angle α can be found by a formula given above.

Now we can proceed to consider a geometric construction to find the directions of polarization of a wave travelling in an arbitrary direction in an anisotropic medium.

The field energy w is given by $8\pi w = D_1^2/\varepsilon_1 + D_2^2/\varepsilon_2 + D_3^2/\varepsilon_3$. Taking new variables $x_k = D_k/\Phi 4\pi w$, we have the equation $x_1^2/\varepsilon_1 + x_2^2/\varepsilon_2 + x_3^2/\varepsilon_3$, which is the equation of an ellipsoid whose semiaxes are the square roots of the principal dielectric constants, or the indices of refraction for waves polarized along a principal axis.

This ellipsoid has been called the *optical indicatrix* or the *index ellipsoid*. Consider the intersection of the plane $x_1s_1 + x_2s_2 + x_3s_3 = 0$ normal to a wave vector s with the ellipsoid. This intersection will be an ellipse. The axes of this ellipse are the directions of polarization of the two waves with the given wave vector.

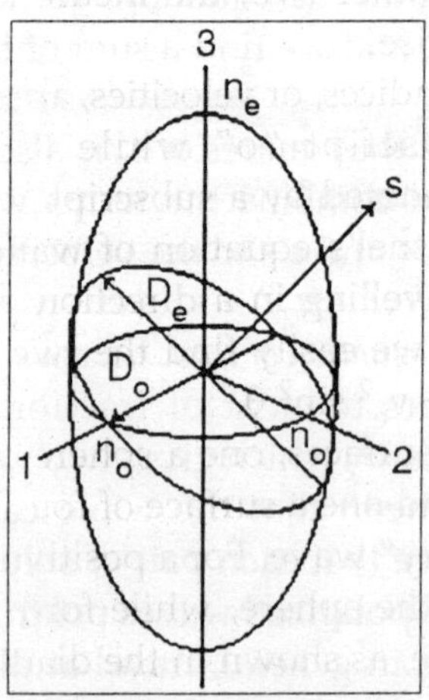

Fig: Index Ellipsoid

It should be clear that once we assume a direction for the wave, the polarization in the perpendicular plane must be carefully chosen so that E and D are coplanar with s, which in general they will not be. This happens for two polarizations that differ by 90°, and the two polarizations will travel with different velocities.

This allows the two polarizations to be separated by refraction, as in a Nicol prism. This is proved by finding the extrema of the vector from the origin to the ellipse, and showing that the result is the same condition that we found above from the combination of Maxwell's equations and the dielectric tensor. Any direction s for which the intersection is a circle is a direction along which a wave may have any polarization, and all polarizations travel at the same velocity.

Such a direction is called an *optic axis* of the crystal. For an ellipsoid with axes all of different lengths, there will be two such directions symmetrically located in the plane of the two axes with the largest and smallest indices on either side of the axis of greatest index. Such media are called *biaxial*.

If two indices are equal, then a plane perpendicular to the axis of the third index cuts the ellipsoid, which is now a spheroid, in a circle. Therefore, this axis is the optic axis, which is the only one. Such media are called *uniaxial*. If the third index is greater than the other two, the medium is called *positive*, and *negative* otherwise.

The two equal indices, or velocities, are called *ordinary* and denoted by a subscript "o" while the third is called *extraordinary* and denoted by a subscript "e."

If we write Fresnel's equation of wave normals for this case, for a wave travelling in a direction making an angle θ with the optic axis, we easily find the two solutions $v^2 = v_o^2$ and $v^2 = v_o^2 \cos^2 \theta + v_e^2 \sin^2 \theta$.

This gives two surfaces, one a sphere of radius v_o for the "ordinary" wave, and one a surface of fourth order, an ovoid, for the "extraordinary" wave. For a positive uniaxial medium, the ovoid is inside the sphere, while for a negative uniaxial medium, it is outside, as shown in the diagram. These figures are not the index ellipsoid.

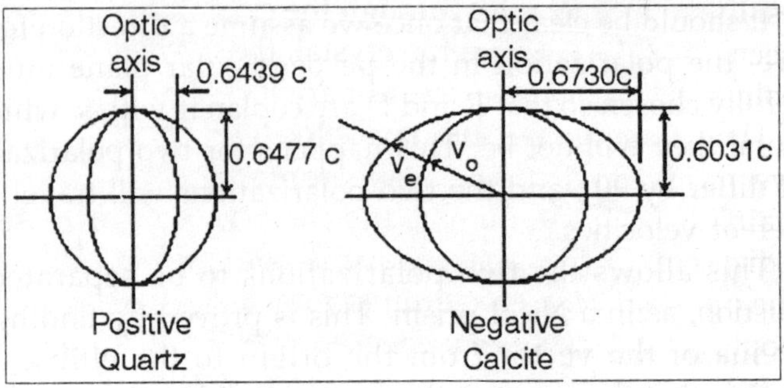

Fig: Uniaxial Wave Normal Surfaces

The distance between the ordinary and extraordinary wave normal surfaces is greatly exaggerated in the diagrams, as can be seen from the actual figures for calcite and quartz that are given.

The extraordinary velocity surface is drawn as an ellipse for convenience, but it is not an ellipse, being blunter at the ends of the major and minor axes than an ellipse. Huygens took it for an ellipse, which indeed it closely resembles, and elementary Optics texts have followed him in this, but this is an error, although it gives the ray direction quite well using Huygen's construction.

This is reminiscent of the Bohr atom, in which erroneous concepts give the correct result. In quartz, the two surfaces do not exactly meet on the optic axis, and this is the source of *optical activity* in quartz, but the effect can be neglected here.

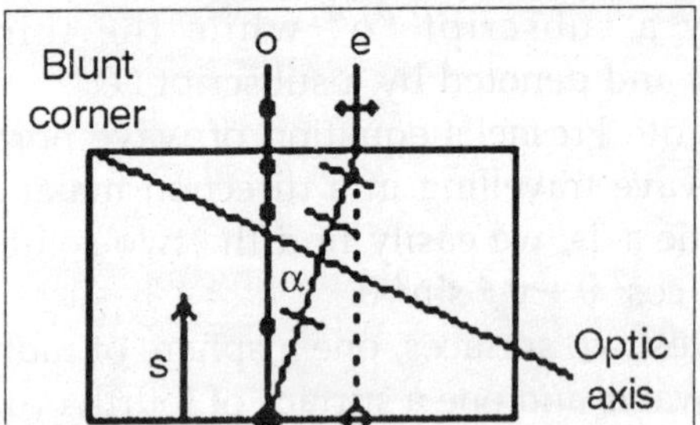

Fig: Double Refraction in Calcite

The optic axis of a cleavage rhomb of calcite passes through a blunt corner making equal angles with the sides and faces. If you look at a dot through the crystal, two images will be seen. When you rotate the crystal, the ordinary image will remain fixed while the extraordinary image rotates around it.

The reason is that the ray velocity of the extraordinary wave is inclined to the wavefronts, lying in the plane containing the wave normal and the optic axis, as shown in the diagram. The ordinary wave is polarized at right angles to the optic axis, the extraordinary ray is polarized parallel to the optic axis, a fact easily checked with a Polaroid filter.

POSTULATES

There are many equivalent ways to state these postulates, and it is entertaining to show that they are all equivalent. Temperature is considered a fundamental notion, abstracted from sensation, that indicates the direction of heat flow and the state of thermal equilibrium. It is often included as a "zeroth" postulate.

In working out the consequences of these postulates, it is very useful to define a function of the state of a system called the *entropy*, with dimensions of energy divided by temperature.

This temperature is a very special one, the *absolute temperature*, for which classical thermodyamics can give a definition, which is a very great accomplishment. Entropy is of great use, and in fact is essential, in engineering and chemical problems. However, what entropy "is," and what are the limits of its application, are not explained at all by the classical theory. English-speaking physicists, in particular, have had a very difficult time understanding it, though they have been able to apply it with great power and ingenuity.

Engineers have been very foggy on entropy, and to all appearances remain so, while engineering students have learned to just follow the rules and think of more pleasant things.

The application of classical mechanics to the explanation of thermodynamics resulted in a complex and disgusting mess that explains nothing, though great efforts were expended along this line.

This is not surprising, since thermodynamics depends sensitively on mechanics at an atomic scale, for which classical mechanics fails spectacularly. Nevertheless, Ludwig Boltzmann and Josiah Willard Gibbs came very close to a reasonable theory, which was completely clarified with the appearance of quantum mechanics.

Gibbs seems to be little remembered these days, but he was the most important physicist that the United States has ever produced, and of world quality.

Since the introduction of quantum mechanics, it has been possible to understand entropy clearly and completely on a very simple basis, so that all the clouds of mystery dissolve into nothingness.

This article will try to give such a short, concise presentation that can easily be comprehended on first principles with very little algebra. The works by Charles Kittel and Peter Atkins should be consulted by anyone interested in these matters. These two investigators understand the whole subject with utter and thorough clarity, and can explain it well.

There is a large body of "scholastic" argument on the foundations of classical thermodyamics and the meaning of entropy that contains nothing of value.

Entropy and the Second Law have been applied outside of thermodynamics to such things as economics and sociology, but these applications reflect only the ignorance of those who propose them, and have no basis. Thermodynamics is very simple, resting only on the existence of quantum-mechanical stationary states that can be counted.

ENTROPY

Thermodynamics applies to systems with a very large number of possible states available to them, and in which transitions between the available states are continual and rapid. Not simply a large number of states is required, but a *very* large number, as argued by Kittel. Very large numbers have properties different from those of merely large numbers, and even their logarithms are large numbers.

Many such systems, and particularly those used as

examples, are composed of numerous subsystems, which may be called *particles*, though we do not mean the structureless dots usually called by this name. Our particles here may be molecules, for example, with internal degrees of freedom in addition to states of motion.

Even simple mass points may have a large number of states available in a macroscopic volume.

Let's take a hydrogen molecule as such a particle, confined in a cube a millimetre on a side. Suppose it has available all the states of kinetic energy less than, say, 4kT, a reasonable thermal excitation energy, where $k = 1.38 \times 10^{-23}$ J/K is Boltzmann's constant. At room temperature, this is about 1.62×10^{-20} J. The quantum number n for a state with energy E is given by $n^2 = 8mL^2E/h^2$. With a mass of 3.32×10^{-27} kg, we find $n = 9.136 \times 10^5$. The total number of states with smaller energy is 1/8 the volume of a sphere of radius n, or 4.00×10^{17} states. Our particle will have at least this many available to it as it jostles with other hydrogen molecules in the cubic millimetre, since there will be about 10^{16} of these. We need have no concern about the rapidity at which transitions between states are made in this system.

Our single particle has more than 10^{17} states available to it. This is already a pretty large number.

There are only about 10^7 seconds in a year, and the universe itself may be only about 10^{18} seconds old. Now we can make another leap and consider the number of states available to the system of hydrogen gas in a cubic millimetre at STP.

This will be the product of the number of states available to one particle raised to the power of the total number of particles, or 10 to the power 17×10^{16}. This is, indeed, a very large number. Stop and consider its breathtaking impudence for a minute. Write it out; 10 to the power of 17 followed by sixteen zeros.

It is so large that multiplying it by, say, 10,000 hardly changes it. Let us denote the number of states available to a system as W, and to a particle as w, so that $W = w^N$, and is a number like the one above. This counting of states is by no means arbitrary or indefinite, but very clearly defined. This is the thing that classical mechanics could not do, and it is fundamental. Matter has a very definite structure, determined by the sizes and properties of atoms, that is beyond sensory perception and the contemplation of fools.

It must be mentioned here that in simply multiplying the number of states of different, but identical, particles we are making an error. In fact, we are counting many states more than once, and have to divide by N!, where N is the number of particles.

We shall take this up later, but at the moment only notice that it will make our estimate of the entropy somewhat too large. Also, hydrogen has rotational energy in addition to translational, and these degrees of freedom contribute to the entropy as well. In this article, we are concerned mainly with the translational degrees of freedom.

The entropy of a system is defined, following Boltzmann, as $S = k \ln W$. This equation is carved on his monument in the Zentralfriedhof zu Wien, and is as remarkable as $E = mc^2$.

The "k" is simply an arbitrary constant that gives S its dimensions, which for temperature in K is more precisely 1.380658×10^{-23} J/K. Kittel represents the entropy as $\sigma = \ln W$ to show that it is essentially a dimensionless counting.

This means that W is an exponential function of S, $W = e^{S/k}$. For our cubic millimetre of hydrogen, $S = k\,(2.3034)(17 \times 10^{16}) = 5.4 \times 10^{-6}$ J/K. Since a mole of hydrogen occupies about 22.4 litre at STP, this gives S = 121 J/mol-K, a not unreasonable result. Of course, this is not an accurate evaluation of the entropy of hydrogen at STP, but shows the principles, and the sizes of numbers, involved.

Note that if we had made an error in estimating W by a factor of 10,000, the answer for the cubic millimetre would only be changed by $k \ln 10{,}000 = 1.27 \times 10^{-22}$ J/K, an unnoticable difference. Entropy is just a logarithmic measure of the number of states available to a system, nothing more mysterious than that. Suppose now that we have two systems, 1 and 2, with available states W_1 and W_2. The number of states available to the system formed by putting 1 and 2 into thermal contact will then be $W = W_1W_2$. Therefore, the entropy of 1 + 2 will be $S = k \ln W = k \ln W_1 + k \ln W_2 = S_1 + S_2$. The entropy, then, is an *additive* property, like mass.

This certainly does not mean that entropy is any kind of substance, as we see from the definition. It cannot be thought of as a fluid (neither can energy, but people persist in thinking this way). When we put 1 and 2 into contact, they may not be in equilibrium, and energy may pass from one to the other, which we call a flow of heat. Let's suppose that system 1 gives up energy dU to system 2.

The loss of dU causes a restriction in the number of available states (some become energetically unreachable), while the gain of dU causes a gain in the number of available states. Since W is a function of U, so is S, and we can express the net change in entropy of 1 + 2 in the form $dS = (\partial S_1/\partial U)(-dU) + (\partial S_2/\partial U)(dU)$.

A more convenient way to write the partial derivatives is to set "S/"U = 1/T. T will have the dimensions of a temperature because of the definition of k, and in fact is just the absolute temperature of the system. Now we have $dS = (1/T_2 - 1/T_1)dU$.

Energy will pass back and forth between the two systems until they have thoroughly explored the available states, and they will come into equilibrium when the two systems jointly have the maximum number of states available to them. Then dS = 0 (since any energy interchange away from the optimum value will lead to fewer available states), and this can be guaranteed for any dU only when $1/T_2 = 1/T_1$, or $T_1 = T_2$.

More than this, if $T_2 > T_1$ initially, dS will be positive for a positive dU, or energy flow from 1 to 2. We have explained not only thermal equilibrium, but the direction of approach to equilibrium by considering numbers of available states. If we know the entropy of a system, we know the temperature of the system. Temperature exists whenever the derivative is well-defined (even for a system of two states only), not just for large systems that can be used as thermal reservoirs, as is important in this example.

We can now briefly indicate the role of time in entropy. A system can be prepared in a non-equilibrium state in which only a limited number of states are available to it. As time passes, more and more states become available, until finally all the states possible given the restrictions on the system (total volume, etc) are available to all the particles.

In terms of entropy, this means that the entropy will increase steadily to a maximum value that is the equilibrium entropy of the system. Think of a system that is created with temperature differences between its parts that only slowly become equalized. Every step towards equalization of temperature gives the particles more and more freedom to occupy different states. The initial temperature differences could be used to drive a heat engine to extract useful work while this takes place. Alternatively, we can just let heat conduction take its course, and lose the possiblility of extracting useful work. The effect on the system is the same (provided we replace the energy extracted) in either case, and the entropy increases to its final maximum value.

Suppose now that the system consists of a large part at a temperature T, and a small part, perhaps one particle, on which we focus our attention. We are interested in the relative probabilities of the small part's being in some state 1 and another single state 2.

To form these states, we have to extract energies E_1 and E_2, respectively, from the large part, which decreases the number of states available to the large part. Let W_1 and W_2 be the available states of the large part in the two cases. Then it is clear that the probability we want is just W_1/W_2. Now $W_1 = e^{S/k}$, and $S_1 = S' - E_1/T$, keeping only the first term in the series, with $dU - -E_1$ and S′ the entropy of the large part before any energy is taken away.

Therefore, $W_1 = e^{S'/k}\exp(-E_1/kT)$, and we have a similar expression for W_2. Therefore, our desired probability is $p_1/p_2 = \exp[-(E_1 - E_2)/kT]$, which is the familiar *Boltzmann factor*.

The reason for the exponential in the Boltzmann factor should be clear. The ratio of probabilities that we want is the ratio of numbers of states, just as if we were working with dice in elementary probability, and the numbers of states are exponential functions of the entropy. Also, our ratio is for individual states; if they come in bunches (degeneracy), then the factors must be multiplied by the numbers of states in each level. The Boltzmann factor is perhaps the most useful result of statistical mechanics, applicable to many problems, and which can even be carried over to problems expressed in classical mechanics, though it is fundamentally a counting of states.

Kittel carries through this argument allowing the number of particles N in the system to be an independent variable as well. The entropy is then a function of N, and its partial derivative with respect to N is $-\mu/kT$, where μ is another very important quantity, the *chemical potential*.

In classical thermodynamics, this quantity is even more mysterious than the entropy, but quantum mechanics gives it a clear meaning, by means of the argument presented here. It is remarkable that the only quantum mechanical result we use directly is simply the existence of stationary states that can be counted. The thermal behaviour of matter is, indeed, conclusive evidence of the existence of atoms. The Second Law is little more than a statement that more probable things are more probable than less probable things.

PARTITION FUNCTION

The probability that state E_i is occupied is $P_i = C\exp(-E_i/kT)$, which is the meaning of the Boltzmann factor.

The constant C can be found by summing over all P_i, which should give unity—the probability that *some* state is occupied. Thus, $1 = C\ \Sigma P_i$, or $C = 1/Z$, where $Z = \Sigma P_i$ is called the *partition function*. Each state of the system is represented in Z by its Boltzmann factor. The internal energy U of the system can be expressed as $U = \Sigma E_i P_i$, the *ensemble average* of E_i. The *ensemble* is an (imaginary) collection of systems, with one in each of the possible states, associated with its corresponding probability. In this case, we are dealing with the Gibbs *canonical ensemble*.

A change in U can be expressed as $dU = \Sigma P_i dE_i + \Sigma E_i dP_i$. The second term is the change in U due to a change in the probabilities, as, for example, caused by a change in temperature. This implies some transitions between particle states if the system is considered as composed of particles. We have already seen this when systems are placed in thermal contact, and heat flows between them. From the definition of temperature, dU = TdS in this case, so we identify this term with TdS.

The first term results from a change in the energy levels without a change in probabilities. For example, if the volume V occupied by the system changes, so do the translational energy levels.

In fact, if the quantum number n remains constant, then the energy of a state is proportional to L^{-2}, if it is confined in a cube of side L. If we consider the state as exerting an outward pressure p_i on each face of its cube, then the work done by the state if L is increased by dL is $3p_iL^2dL = p_idV$, so $dE_i = -p_idV$. The term ΣP_idE_i is the ensemble average of the work done, or -pdV. Now we have dU = -pdV + TdS, which is called the *thermodynamic identity.*

Changes in U can take place due to changes in occupation of states, or by changes in the energies of the states, and these changes are called heat and work, respectively.

We assume that the changes associated with pdV take place with no change of entropy—that is, they are "reversible" in the usual terminology of thermodynamics. Work expressed as pdV is only one kind of work that can be done by a system; the energy of a state may depend on other parametres besides the volume, such as the magnetic field, which can be treated analogously.

We have just seen that $(\partial U/\partial V)_S = -p$. This is merely another way to write that the change in energy dU caused by a change in volume dV at constant entropy S is the negative of the pressure, or dU – –pdV (S = constant), which we have just discussed.

A change in S can be expressed in terms of partial derivatives by $dS = (\partial S/\partial U)_V\ dU + (\partial S/\partial V)_U\ dV$, where S is considered as a function of U and V, as we have assumed above. Now set dS = 0, divide by dV, and replace dU/dV by $(\partial U/\partial V)_S$ by -p. Since $(\partial S/''U)_V = 1/T$, we then have $(\partial S/''V)_U =$

-p/T. One might be tempted to use this expression as a definition of p, as we defined T by a partial derivative of S, but that would not be legal, since p is already defined in terms of the forces exerted by the system.

What we have to do is show that this externally defined p also appears in the partial derivative expression, and we have just done that.

Just as in the case of temperature, we can now show that $p_1 = p_2$ in equilibrium, and that if $p_1 > p_2$, V_1 increases at the expense of V_2, so that system 1 does work on system 2. The systems must also be in thermal equilibrium for this reasoning to be valid. If they are not, we must have recourse to mechanical arguments, since dS will be zero in any case.

Since $P_i = \exp(-E_i/kT)/Z$, $\ln P_i = -E_i/kT - \ln Z$, or $E_i = -kT \ln P_i - kT\ln Z$. Since $\Sigma P_i = 1$, $\Sigma dP_i = 0$, so multiplying the previous equation by dP_i and summing over i, we find $\Sigma E_i dP_i = -kT \Sigma dP_i \ln P_i$. The sum on the left is, as we have seen, just TdS, and $d[\Sigma P_i \ln P_i] = \Sigma\, dP_i \ln P_i$ since $\Sigma dP_i = 0$, so $dS = d[-kT\, \Sigma P_i \ln P_i]$, and thus $S = -kT\, \Sigma P_i \ln P_i$.

The constant of integration is zero, as we can see by considering a system occupying a single state. This is Boltzmann's expression for the entropy in terms of the probabilities.

Now we go back to the first equation in the preceding paragraph, where $-kT \ln Z = E_i + kT \ln P_i$. Multiplying by P_i and summing, which is taking the ensemble average, we find $-kT \ln Z = U - TS$, using the expression for S that we have just derived.

The function U – TS is the *free energy* F. It has the property that $dF = dU - TdS - SdT = -pdV - SdT$, so that $dF = 0$ in a process where $dV = dT = 0$, or an isothermal process at constant volume, such as a chemical reaction taking place in a fixed volume at a fixed temperature.

This means that F is an extremum (usually a minimum) at equilibrium, and can be used as a criterion of equilibrium in an isothermal process as entropy is used in a process at constant energy. The fact that $F = -kT \ln Z$ makes the partition function very useful for finding the thermodynamic functions of a system. From F, we can find $p = -(\partial F/\partial V)_T$, and $S = -(\partial F/\partial T)_V$ by differentiation, and then U = F + TS, and so on. These properties come out as functions of T and V, which is much more convenient than the expressions in terms of U and V that we have found up to this point.

As an example, consider a linear harmonic oscillator whose states have energies (n + 1/2)hf, where n is the quantum number, f the frequency, and h Planck's constant. We have included the zero-point energy hf/2.

Then, $Z = \Sigma \exp[-(n + 1/2)(hf/kT)] = \Sigma \exp[-(n + 1/2)x]$, where x = hf/kT. This series can be summed to $Z = \exp(-x/2)/(1 - e^{-x}) = e^{x/2}/(e^x - 1)$. Therefore, $F = -kT[(x/2) - \ln(e^x - 1) = -hf/2 + kT \ln(e^x - 1)$. The entropy is $S = -k \ln (e^x - 1) + (hf/T)e^x/(e^x - 1)$.

Finally, $U = F + TS = hf/(1 - e^{-x}) - hf/2$. For kT >> hf, U → kT, while for kT << hf, U → hf/2, which are the correct limits. The specific heat is $C = dU/dT = kx^2e^{-x}/(1 - e^{-x})^2$, the Einstein expression. For kT >> hf (large x) C → k. The Einstein temperature is $T_E = hf/k$, a measure of the quantum size. When $T = T_E$, then C = 0.92k, only 8% below its high-temperature limit. When $T = T_E/10$, C = 0.0045k, and the oscillator is "frozen out." This has an important application to the vibration of molecules. The Einstein temperature of O_2 is 2260K, and of H_2 6300K. The vibration of hydrogen is frozen out at room temperature, while that of oxygen is only a little excited.

If the zero-point energy is omitted, then $Z = \Sigma e^{-nx} = 1/(1 - e^{-x}$. We find $F = kT \ln (1 - e^{-x})$ and $S = -k \ln (1 - e^{-x}) + (hf/T)/(e^x - 1)$. Then, $U = F + TS = hf/(e^x - 1)$. This is appropriate for photons, which are not conserved and so cannot have a zero-point energy. We recognize here the Bose-Einstein distribution for zero chemical potential, which appears in the theory of black-body radiation. The spectrum of black-body radiation (the energy density) is this quantity times the number of modes (oscillators) in the frequency interval df.

A two-state system with a ground state of energy 0 and an excited state of energy E can also be easily handled, since the partition function is the simple expression $Z = 1 + e^{-E/kT}$. Then, $U = Ee^{-x}/(1 + e^{-x}) = E/(e^x + 1)$, where x = E/kT. The specific heat is $C = kx^2e^x/(e^x + 1)^2$. This is the same as for the harmonic oscillator, except that a +1 replaces the -1 in the denominator, which makes a considerable difference. For $x \to \infty$ (low temperatures), $C \to 0$, while for $x \to 0$ (high temperatures), C also approaches zero.

The value of x for the maximum C can be found by setting the derivative of C with respect to x equal to zero. The result is the solution of $2/x = \tanh(x/2)$, which is about $x = 2.40$. The peak of the C versus T curve is at $T = 2.40E/k$, from which E can be estimated experimentally. This peak is called a *Schottky anomaly*.

The entropy can be found in the usual way from $F = -kT \ln (1 + e^{-x})$. If we set $E = 1$ for simplicity, we find $S = -k[U \ln U + (1 - U)\ln(1 - U)]$. This is zero for $U = 0$ and $U = 1$ (note that $\lim(x \to 0)\ x \ln x = 0$), and symmetrical about a maximum at $U = 1/2$, for which $S = k \ln 2$.

The temperature is given by $1/T = dS/dU = k \ln[(1 - U)/U]$. We must replace k by k/E if the temperature is to come out in K. For $U = 0$, or the upper state unoccupied, we find $T = 0$, which is not unexpected. For $U < 1/2$, the temperature is positive, approaching $\int$ for $U = 1/2$, where the two states are equally occupied on the average.

For still greater U, the temperature becomes *negative*, approaching 0 from below as $U \to E$, when the lower state is unoccupied. Negative temperatures are higher than positive temperatures, with -0K the highest temperature of all. Whenever the temperature is negative, the upper state is more populated than the lower, and there is a *population inversion*.

If we put systems with temperatures +T and -T in contact, the result at equilibrium will be $T = \infty$, not zero. This is quite reasonable, since in the combined system the upper and lower levels will be equally occupied, which corresponds to infinite temperature. Population inversions are created in lasers so that the system will radiate spontaneously.

THE IDEAL GAS

An ideal gas is a system consisting of N identical particles in a volume V. The particles are assumed to move entirely independently. Of course, in any real case they collide very frequently, and these collisions are necessary to permit the system to occupy all of its available states. Nevertheless, the particles feel each other's presence very rarely, and are *almost* independent. We shall totally neglect the collisions, and trust that the gas can reach equilibrium quickly when disturbed. We will consider only the translational motion of the particles, and not any internal degrees of freedom they may have. This restriction is purely for simplicity.

Let's begin by considering one molecule in a volume V, thought of as a cube of side L, such that $L^3 = V$.

The states are labelled by three quantum numbers, and the energy E depends only on the sum of squares n^2 of the quantum numbers. In fact, $E = h^2n^2/8mL^2$, where m is the mass of the particle. In n-space, where the quantum numbers are plotted along three orthogonal axes, there is one state per unit volume, and only points in the first octant correspond to different states. This means that the number of states in (n,n+dn) is $D(n)dn = \pi^2n^2dn/2$. Now it is easy to find Z by replacing the sum with an integral. Setting $x = (n/L)(h/8mkT)^{1/2}$, we have $Z = [L(2mkT/h^2)]^3 (\pi/2) \int(0, \infty) x^2\exp(-x^2)dx$.

The value of the definite integral is $\pi^{1/2}/4$, so $Z = V[2\pi mkT/h^2]^{3/2}$. This is the partition function for a single particle.

The energy can be found from Z by the formula $U = kT^2\partial(\ln Z)/\partial T$ (this is just $-\partial F/\partial T$). The only temperature-dependent part of ln Z is $T^{3/2}$, so we find $U = 3kT/2$. This is an expression of the famous *principle of equipartition*, which states that to each degree of mechanical freedom an energy kT/2 corresponds in thermal equilibrium.

In an ideal gas we have three degrees of translational freedom. A harmonic oscillator has two degrees of freedom, one for the kinetic energy and one for the potential energy, and we have already seen that its average energy is kT at a high enough temperature.

The pressure is given by $-\partial F/\partial V$, or $p = kT\,\partial(\ln Z)/\partial V$. The only volume-dependent part of Z is simply ln V, so we find $p = kT/V$ or, $pV = kT$, the ideal gas law for a single particle.

If we have N particles instead of just one, it would not take much argument to convince one that $U = 3NkT/2$, and

$pV = NkT$, instead. These are, indeed, correct results, and would be obtained if we simply raised the single-particle Z to the Nth power.

It happens, however, that if we do this, the free energy and the entropy do not come out correctly! The problem is quantum-mechanical.

In the microscopic world, two electrons, or two hydrogen atoms, are not just similar, they are identical. If we have a state containing any number of them, interchanging any pair gives back exactly the same state, not a different one. That is, the number of states calculated on the basis that the particles are all distinguishable (raising Z to the power N, for example) is too large by a factor N!.

This is the number of ways we can permute the particles in the state and get back the same state. It is actually a little worse than this. The states of particles like electrons, that have half-integral spins, must be antisymmetric on the exchange of any two particles (i.e., it must change sign), while states of particles of integral spin must be symmetric under exchange. We will not go into the complexities of state-counting that these considerations require, but stick with the simple N! factor.

The partition function for N particles is then $Z = (V^N/N!)[2\pi mkT/h^2]^{3N/2}$. In handling the N!, the Stirling approximation $\ln N! = N \ln N - N$ will be accurate enough. The internal energy U and the pressure p come out as above, but F and $S = -\partial F/\partial T$ are different.

Most important is the expression for the translational entropy, $S = (3Nk/2)\ln(2\pi mk/h^2) + (5Nk/2) + Nk \ln(V/N) + (3Nk/2)\ln T$, which differs from the naive result by $-k(N \ln N - N)$, the contribution of the N!. This is the famous *Sakur-Tetrode* (Tetrode has three syllables) equation.

For the cubic millimetre of hydrogen we used as an example above, the first two terms, the constant terms, contribute $+8.842 \times 10^{-6}$ J/K to the entropy; the third term gives -7.63×10^{-6} J/K. Part is volume-dependent, but the rest is constant. The final term, the temperature dependence, adds 1.180×10^{-6} J/K, for a total of 2.392×10^{-6} J/K.

This is a much more accurate value than our early rough estimate, but it is of the same order of magnitude. The Sakur-Tetrode equation has been thoroughly confirmed by experiment. For a monatomic gas, such as helium or argon, it gives the total entropy.

We have seen that the usual ideal gas constant $R = N_A k$ = 8.31451 J/mol-K. Note that this is for a gram-mole, for which $N_A = 6.02 \times 10^{23}$. It is sometimes given for a kg-mole, a unit which is never used by the sane. The Boltzmann constant is just the gas constant per molecule, and is the same constant used to get entropy from probability in S = k ln W.

Since 1 cal = 4.186 J, R = 1.986 cal/mol-K, quite close to the easily-remembered round value of 2 cal/mol-K. Measurements on ideal gases is one way to determine k experimentally. As a practical matter, most gases under normal conditions are very well approximated as ideal gases, consisting of independent, classical particles, and their thermodynamic properties are easily determined.

Linear molecules have an additional two rotational degrees of freedom, and nonlinear molecules have three. Rotational energies are small, so they are not frozen out at usual temperatures as vibration.

The prinᶜiple of equipartition then applies, so the internal energy is increased by RT and 3RT/2 per mole, respectively, giving U = 5RT/2 and U = 3RT for the two cases.

The specific heat at constant volume is dU/dT, so C_V = 3R/2 for a monatomic gas, 5R/2 for a gas of linear molecules (such as a diatomic gas) and 3R for a gas of nonlinear molecules (such as a polyatomic gas). These predictions are well confirmed by experiment. An *adiabatic* process is one that takes place without heat exchange. If it is *reversible*, then it will be *isentropic*.

Such processes are usually fairly rapid ones, such as the compressions involved in the propagation of sound, or the convective rise of bodies of air in the atmosphere, and are of great practical importance. For a monatomic gas, S = (3R/2)ln T + R ln V + constant terms J/mol-K. Therefore, S = constant implies that $T^{3R/2}V^R$ = constant, or $VT^{3/2}$ = constant. Since pV

= RT, this can be cast into the form $V(pV/R)^{3/2}$ = constant, or $p^{3/2}V^{5/2}$ = constant.

As $C_V = 3R/2$ and $C_p = C_V + R = 5R/2$ (this latter relation is proved in the Appendix; C_p is the specific heat at constant pressure, when the gas must be allowed to expand and do work). In view of this, an isentropic process for an ideal monatomic gas is characterized by $p^{Cv}V^{Cp}$ = constant, or $pV^{Cp/Cv}$ = constant. It is shown in the Appendix that this is a general relation.

The ratio of specific heats Cp/Cv is usually denoted by γ, so finally we have the familiar pV^{γ} = constant. For a monatomic gas, $\gamma = 5/3 = 1.667$, for a gas of linear molecules $\gamma = 7/5 = 1.400$ and for a gas of nonlinear molecules, $\gamma = 8/6 = 1.333$. The ratio of specific heats can be determined quite accurately by measuring the speed of sound, and these predictions are fully confirmed.

DIFFERENCE IN THE SPECIFIC HEATS

We begin with the thermodynamic identity, $dU = TdS - pdV$, and consider one mole of substance. In a change of temperature dT at constant pressure, $dU = C_V dT$ for an ideal gas, while the heat absorbed $TdS = C_p dT$.

Hence, $C_p = C_V + p(\partial V/\partial T)_p$, where the partial derivative is the ratio of dV to dT at constant p. Since the equation of state of one mole of an ideal gas is $pV = RT$, $p(\partial V/\partial T)_p = R$. Therefore, $C_p = C_V + R$. Q.E.D.

In general, the internal energy can depend on the volume as well as on the temperature. In fact, $(\partial U/\partial V)_T = T(\partial p/\partial T)_V - p$, which is zero for an ideal gas. In the more general case, $C_p - C_V = -T(\partial V/\partial T)^2(\partial p/\partial V)_T$. The necessary derivatives can be found from the equation of state.

RELATION BETWEEN P AND V IN AN ISENTROPIC PROCESS

The thermodynamic identity is again the starting point, $dU = TdS - pdV$. In an isentropic process, $dS = 0$, so $0 = C_V dT + (RT/V)dV$, where we have used the equation of state for one mole of an ideal gas, and have made the same assumption about the internal energy as above. Then $C_V(dT/T) + R(dV/V) = 0$. This integrates easily to $C_V \ln T + R \ln V$ = constant, or $T^{C_V}V^R$ = constant, or $T^{C_V/R}V$ = constant.

From the equation of state, $T = pV/R$, so $p^{Cv/R}V^{Cv/R+1}$ = constant. Raising to the power R/Cv, we find pV^γ = constant, where $\gamma = (Cv/R + 1)/Cv = Cp/Cv$. this result is exact for an ideal gas only. Q.E.D.

RELATION BETWEEN ADIABATIC AND ISOTHERMAL BULK MODULI

The bulk modulus is defined by $k = -V(dp/dV)$, where an increase in pressure dp causes a decrease in volume -dV. If the temperature is held constant, then $pV = RT$ = constant, and the isothermal bulk modulus is $k_T = p$ for an ideal gas. If the process is adiabatic, then pV^γ = constant.

The derivative with respect to p is $V^\gamma + p\gamma V^{\gamma-1}(dV/dp)_s$. The adiabatic (isentropic) bulk modulus is then $k_s = \gamma p$. The speed of sound c is given by $c^2 = k/\rho$. For an ideal gas, this becomes $c^2 = \gamma p/\rho = \gamma RT/M$. The adiabatic process is "stiffer" than the isothermal. Although we have proved it only for an ideal gas, the relation $\kappa_s = \gamma k_T$ is true in general.

THE GAS CONSTANT IN OTHER UNITS

We gave two values for the gas constant above, 8.314 J/mol-K and 2 cal/mol-K, which are convenient in some problems, but not in all. The gas constant for one mole in any units can be found from $R = pV/T$, where corresponding values for p, V and T are inserted. For example, p = 1 atm, V = 22.4 l and T = 273.15K are well-known corresponding values, which give R = (1)(22.4)/(273.15) = 0.082006 l-atm/mol-K.

Note that 0.0820 is a value good to better than 1 part in 10,000, and not too hard to remember.

If you are converting to U.S. Engineering Units, you need to know that a pound-mole is 453.59237 times more than a gram-mole, and a °R (rankine) is 9/5 of a kelvin (0°C is 491.67R). At STP the pound-molar volume is 358.8 cubic feet, so R = 0.73 cuft-atm/lb-mol-R. Since 1 atm is 14.696 psi, R = 10.73 psi-cuft/lb-mol-R as well.

TWO-COMPONENT SYSTEMS

The simplest imaginable multi-component dynamical system consists of two point mass objects which are both constrained to move along the same straight-line. Let x_1 be the displacement of the first object, whose mass is m_1.

Likewise, let x_2 be the displacement of the second object, whose mass is m_2. Suppose that the first object exerts a force f_{21} on the second object, whereas the second object exerts a force f_{12} on the first. From Newton's third law of motion, we have

$$f_{12} = -f_{21}.$$

Suppose, finally, that the first object is subject to an external force (i.e., a force which originates outside the system) F_1, whilst the second object is subject to an external force F_2. Applying Newton's second law of motion to each object in turn, we obtain

$$m_1 \ddot{x}_1 = f_{12} + F_1,$$
$$m_2 \ddot{x}_2 = f_{21} + F_2.$$

Here, is a convenient shorthand for d/dt. Likewise, means d^2/dt^2.

At this point, it is helpful to introduce the concept of the centre of mass. The centre of mass is an imaginary point whose displacement x_{cm} is defined to be the mass weighted average of the displacements of the two objects which constitute the system. In other words,

$$x_{cm} = \frac{m_1 x_1 + m_2 x_2}{m_1 + m_2}$$

Thus, if the two masses are equal then the centre of mass lies half way between them, if the second mass is three times larger than the first then the centre of mass lies three-quarters of the way along the line linking the first and second masses, respectively, if the second mass is much larger than the first then the centre of mass is almost coincident with the second mass, and so on.

Summing Eqs. and, and then making use of Eqs. and, we obtain

$$m_1 \ddot{x}_1 + m_2 \ddot{x}_2 = (m_1 + m_2) \ddot{x}_{cm} = F_1 + F_2.$$

Note that the internal forces, f_{12} and f_{21}, have canceled out. The physical significance of this equation becomes clearer if we write it in the following form.

$$M\ddot{x}_{cm} = F,$$

where $M = m_1 + m_2$ is the total mass of the system, and $F = F_1 + F_2$ is the net external force acting on the system. Thus, the motion of the centre of mass is equivalent to that which would occur if all the mass contained in the system were collected at the centre of mass, and this conglomerate mass were then acted upon by the net external force. In general, this suggests that the motion of the centre of mass is simpler than the motions of the component masses, m_1 and m_2.

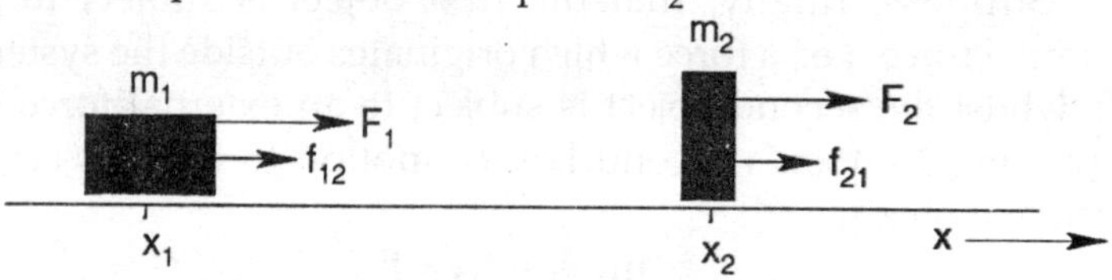

Fig: A 1-dimensional Dynamical System
Consisting of two point mass Objects

This is particularly the case if the internal forces, f_{12} and f_{21}, are complicated in nature.

Suppose that there are no external forces acting on the system (i.e., $F_1 = F_2 = 0$), or, equivalently, suppose that the sum of all the external forces is zero (i.e., $F = F_1 + F_2 = 0$). In this case, according to Eq., the motion of the centre of mass is governed by Newton's first law of motion. i.e., it consists of uniform motion in a straight-line. Hence, in the absence of a net external force, the motion of the centre of mass is almost certainly far simpler than that of the component masses.

Now, the velocity of the centre of mass is written,

$$v_{cm} = \dot{x}_{cm} = \frac{m_1\dot{x}_1 + m_2\dot{x}_2}{m_1 + m_2}.$$

We have seen that in the absence of external forces v_{cm} is a constant of the motion (i.e., the centre of mass does not accelerate). It follows that, in this case,

$$m_1\dot{x}_1 + \dot{x}_2\dot{x}_2$$

is also a constant of the motion. Momentum is defined as the product of mass and velocity. Hence, the momentum of the first mass is written $p_1 = m_1 \dot{x}_1$, whereas the momentum of the second mass takes the form $p_2 = m_2 \dot{x}_1$. It follows that the above expression corresponds to the total momentum of the system.

$$P = p_1 + p_2.$$

Thus, the total momentum is a conserved quantity—provided there is no net external force acting on the system. This is true irrespective of the nature of the internal forces. More generally, Eq. can be written

$$\frac{dP}{dt} = F.$$

In other words, the time derivative of the total momentum is equal to the net external force acting on the system—this is just Newton's second law of motion applied to the system as a whole.

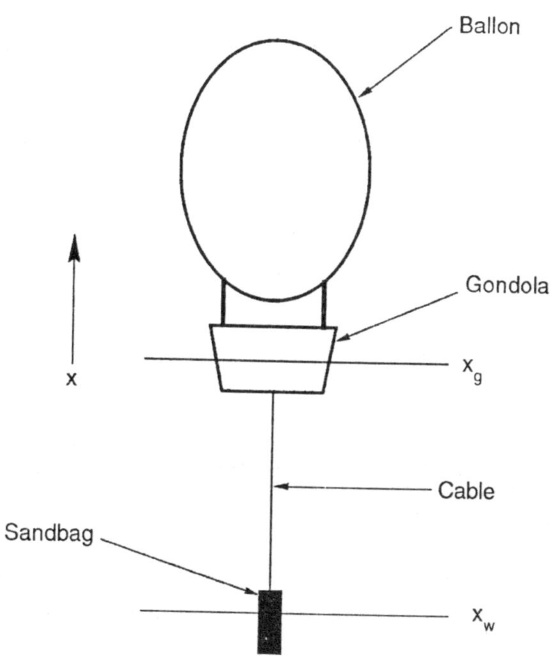

Fig: An Example Two-component System

Let us now try to apply some of the concepts discussed

above. Consider the simple two-component system shown. A gondola of mass m_g hangs from a hot-air balloon whose mass is negligible compared to that of the gondola. A sandbag of mass m_w is suspended from the gondola by means of a light inextensible cable.

The system is in equilibrium. Suppose, for the sake of consistency with our other examples, that the x-axis runs vertically upwards. Let x_g be the height of the gondola, and x_w the height of the sandbag. Suppose that the upper end of the cable is attached to a winch inside the gondola, and that this winch is used to slowly shorten the cable, so that the sandbag is lifted upwards a distance Δx_w. The question is this. does the height of the gondola also change as the cable is reeled in? If so, by how much?

Let us identify all of the forces acting on the system. The internal forces are the upward force exerted by the gondola on the sandbag, and the downward force exerted by the sandbag on the gondola. These forces are transmitted via the cable, and are equal and opposite (by Newton's third law of motion). The external forces are the net downward force due to the combined weight of the gondola and the sandbag, and the upward force due to the buoyancy of the balloon. Since the system is in equilibrium, these forces are equal and opposite (it is assumed that the cable is reeled in sufficiently slowly that the equilibrium is not upset).

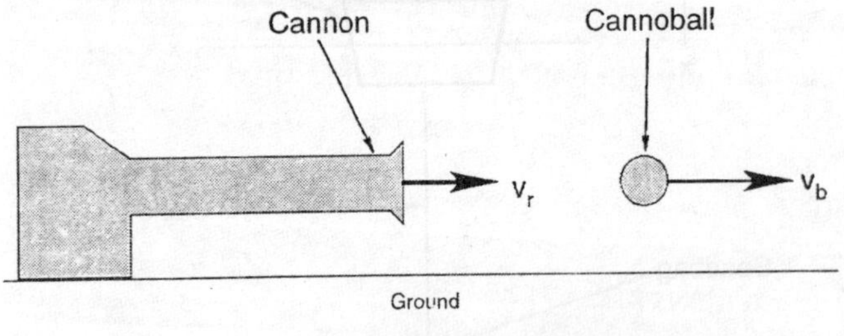

Fig. Another Example Two-component System

Hence, there is zero net external force acting on the system. It follows, from the previous discussion, that the centre of mass of the system is subject to Newton's first law. In particular, since the centre of mass is clearly stationary before the winch is turned on, it must remain stationary both during and after the period in which the winch is operated. Hence, the height of the centre of mass,

$$x_{cm} = \frac{m_g x_g + m_w x_w}{m_g + m_w},$$

is a conserved quantity. Suppose that the operation of the winch causes the height of the sandbag to change by Äx$_w$, and that of the gondola to simultaneously change by Äx$_g$. If x_{cm} is a conserved quantity, then we must have

$$0 = m_g\ \Delta x_g + m_w\ Äx_w,$$

or

$$\Delta x_g = \frac{m_w}{m_g} \Delta x_w .$$

Thus, if the winch is used to raise the sandbag a distance Δx_w then the gondola is simultaneously pulled downwards a distance $(m_w = m_g)\ \Delta x_w$. It is clear that we could use a suspended sandbag as a mechanism for adjusting a hot-air balloon's altitude. the balloon descends as the sandbag is raised, and ascends as it is lowered.

Suppose that a cannon of mass M propels a cannonball of mass m horizontally with velocity v_b. What is the recoil velocity v_r of the cannon? Let us first identify all of the forces acting on the system. The internal forces are the force exerted by the cannon on the cannonball, as the cannon is Äred, and the equal and opposite force exerted by the cannonball on the cannon. These forces are extremely large, but only last for a short instance in time. in physics, we call these impulsive forces. There are no external forces acting in the horizontal direction (which is the only direction that we are considering in this example). It follows that the total (horizontal) momentum P of the system is a conserved quantity. Prior to the Firing of the cannon, the total momentum is zero (since momentum is mass times velocity, and nothing is initially moving). After the cannon is Äred, the total momentum of the system takes the form,

$$P = mv_b + Mv_r.$$

Since P is a conserved quantity, we can set P = 0. Hence,

$$v_r = -\frac{m}{M} v_b .$$

Thus, the recoil velocity of the cannon is in the opposite direction to the velocity of the cannonball (hence, the minus sign in the above equation), and is of magnitude (m=M) v_b.

Of course, if the cannon is far more massive that the cannonball (i.e., M >> m), which is usually the case, then the recoil velocity of the cannon is far smaller in magnitude than the velocity of the cannonball. Note, however, that the momentum of the cannon is equal in magnitude to that of the cannonball. It follows that it takes the same effort (i.e., force applied for a certain period of time) to slow down and stop the cannon as it does to slow down and stop the cannonball.

MULTI-COMPONENT SYSTEMS

Consider a system of N mutually interacting point mass objects which move in 3-dimensions. Let the ith object, whose mass is m_i, be located at vector displacement r_i. Suppose that this object exerts a force f_{ji} on the jth object. By Newton's third law of motion, the force f_{ij} exerted by the jth object on the ith is given by,

$$f_{ij} = -f_{ji}.$$

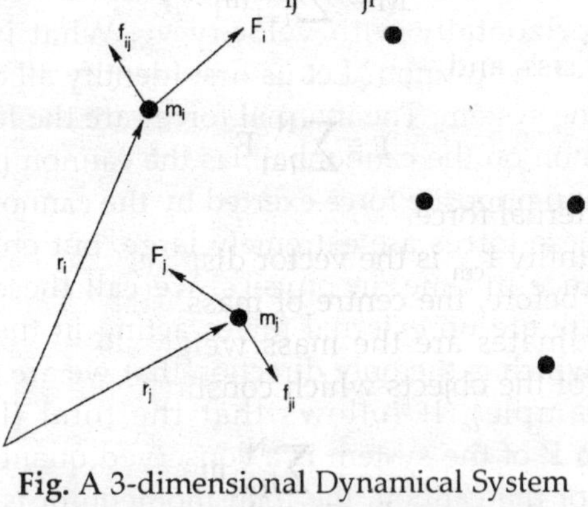

Fig. A 3-dimensional Dynamical System Consisting of Many Point Mass Objects.

Finally, suppose that the ith object is subject to an external force F_i. Newton's second law of motion applied to the ith object yields,

$$m_i \ddot{r}_i = \sum_{j=1,N}^{j \neq i} f_{ij} + F_i$$

Note that the summation on the right-hand side of the above equation excludes the case j = i, since the ith object cannot exert a force on itself. Let us now take the above equation and sum it over all objects.

We obtain,

$$\sum_{i=1,N} m_i \ddot{r}_i \sum_{i=1,N}^{j \neq i} f_{ij} + \sum_{i=1,N} F_i .$$

Consider the sum over all internal forces. i.e., the first term on the right-hand side.

Each element of this sum—f_{ij}, say—can be paired with another element— f_{ji}, in this case—which is equal and opposite. In other words, the elements of the sum all cancel out in pairs. Thus, the net value of the sum is zero. It follows that the above equation can be written,

$$M \ddot{r}_{cm} = F,$$

where,

$$M = \sum_{i=1}^{N} m_i$$

is the total mass, and,

$$F = \sum_{i=1}^{N} F_i$$

is the net external force.

The quantity r_{cm} is the vector displacement of the centre of mass. As before, the centre of mass is an imaginary point whose coordinates are the mass weighted averages of the coordinates of the objects which constitute the system. Thus,

$$r_{cm} = \frac{\sum_{i=1}^{N} m_i r_i}{\sum_{i=1}^{N} m_i}$$

Index

H

I

J

K

M

O

P

R